Rul[illegible]
fo[illegible]
D[illegible]ibutors 2017

Rules and Guidance for Pharmaceutical Distributors 2017

Compiled by the Inspection, Enforcement and Standards Division of MHRA

 Medicines & Healthcare products Regulatory Agency

Published by Pharmaceutical Press

66–68 East Smithfield, London E1W 1AW, UK

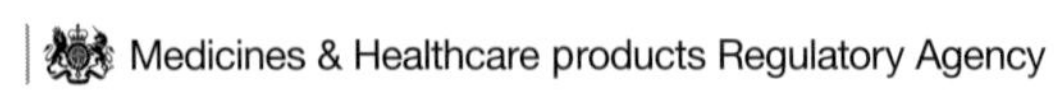

MHRA, a centre of the Medicines and Healthcare products Regulatory Agency
151 Buckingham Palace Road
Victoria
London SW1W 9SZ

Designed and published by Pharmaceutical Press 2017

(PhP) is a trade mark of Pharmaceutical Press

Pharmaceutical Press is the publishing division of the Royal Pharmaceutical Society

First edition published in 2007, second edition 2014, third edition 2015, fourth edition 2017

Typeset by OKS Group, Chennai, India
Printed in Great Britain by TJ International, Padstow, Cornwall

ISBN 978 0 85711 286 6 (print)
ISBN 978 0 85711 287 3 (epdf)
ISBN 978 0 85711 288 0 (ePub)
ISBN 978 0 85711 289 7 (mobi)

A catalogue record for this book is available from the British Library.

Contents

EU referendum xi

Preface to the 2017 edition xii

Acknowledgements xvii

Feedback xviii

Introduction xix

Glossary of Legislation xxii

MHRA 1

1 MHRA: Licensing, Inspection and Enforcement for Medicines 3

Overview of the Medicines and Healthcare products Regulatory Agency Group 3
Overview of MHRA 5
Inspectorate 8
The Data Processing Group 10
Defective Medicines Report Centre (DMRC) 12
Enforcement Group 13
Compliance Management and Inspection Action Group 14
Advice 19

Wholesale Distribution of Medicines 21

2 EU Legislation on Wholesale Distribution 23

Directive 2001/83/EC, as Amended, Title VII, Wholesale Distribution and Brokering Medicines 25

Directive 2001/83/EC of the European Parliament and of the Council of 6 November 2001 on the Community code relating to medicinal products for human use as amended 25
Title VII: Wholesale Distribution and Brokering of Medicinal Products 25
Commission Delegated Regulation (EU) 2016/161 of 2 October 2015 supplementing Directive 2001/83/EC of the European Parliament and of the Council by laying down detailed rules for the safety features appearing on the packaging of medicinal products for human use 32
Chapter I: Subject Matter and Definitions 33
Chapter II: Technical Specifications of the Unique Identifier 34
Chapter III: General Provisions on the Verification of the Safety Features 38
Chapter IV: Modalities of Verification of the Safety Features and Decommissioning of the Unique Identifier by Manufacturers 39
Chapter V: Modalities of Verification of the Safety Features and Decommissioning of the Unique Identifier by Wholesalers 41
Chapter VI: Modalities of Verification of the Safety Features and Decommissioning of the Unique Identifier by Persons Authorised or Entitled to Supply Medicinal Products to the Public 43
Chapter VII: Establishment, Management and Accessibility of the Repositories System 46
Chapter VIII: Obligations of Marketing Authorisation Holders, Parallel Importers and Parallel Distributors 55
Chapter IX: Obligations of the National Competent Authorities 56
Chapter X: Lists of Derogations and Notifications to the Commission 57
Chapter XI: Transitional Measures and Entry Into Force 58
Annex I 60
Annex II 61
Annex III 62
Annex IV 63

3 UK Legislation on Wholesale Distribution 64

The Human Medicines Regulations 2012 [SI 2012/1916] 64

4 EU Guidelines on Good Distribution Practice of Medicinal Products for Human Use (2013/C 343/01) 82

Guidelines on Good Distribution Practice of Medicinal Products for Human Use (2013/C 343/01) 83

Introduction 83
Chapter 1 – Quality Management 84
Chapter 2 – Personnel 86
Chapter 3 – Premises and Equipment 88
Chapter 4 – Documentation 92
Chapter 5 – Operations 93
Chapter 6 – Complaints, Returns, Suspected Falsified Medicinal Products and Medicinal Product Recalls 97
Chapter 7 – Outsourced Activities 100
Chapter 8 – Self-inspections 101
Chapter 9 – Transportation 101
Chapter 10 – Specific Provisions for Brokers 104
Chapter 11 – Final Provisions 105
Annex 106
European Commission Q&A on GDP guidelines 108

5 UK Guidance on Wholesale Distribution Practice 113

The Application and Inspection Process: "What to Expect" 114
Risk-based Inspection Programme 118
Conditions of Holding a Wholesale Dealer's Licence 121
Appointment and Duties of the Responsible Person 128
The Responsible Person Gold Standard 130
Quality Management 135
Controls on Certain Medicinal Products 137
Sourcing and Exporting Medicinal Products – Non-EEA Countries 145
Temperature Control and Monitoring 155
Handling Returns 164
Short-term Storage of Ambient and Refrigerated Medicinal Products – Requirements for a Wholesale Dealer's Licence 166
Sales Representative Samples 167
Qualification of Customers and Suppliers 168
Falsified Medicines 171
Regulatory Action 173
Diverted Medicines 173
Parallel Importation 173
Parallel Distribution 175
Continued Supply 175
Matters Relating to Unlicensed Medicines 177
Reporting Adverse Reactions 180
Product Recall/Withdrawal 180
Data Integrity 185

Brokering of Medicines 191

6 EU Legislation on Brokering Medicines 193

Directive 2001/83/EC, as Amended, Title VII, Wholesale Distribution and Brokering Medicines 193
Directive 2001/83/EC of the European Parliament and of the Council of 6 November 2001 on the Community code relating to medicinal products for human use as amended 193
Title VII: Wholesale Distribution and Brokering of Medicinal Products 193

7 UK Legislation on Brokering Medicines 196

The Human Medicines Regulations 2012 [SI 2012/1916] 196

8 EU Guidelines on Good Distribution Practice of Medicinal Products for Human Use (2013/C 343/01) 202

Guidelines on Good Distribution Practice of Medicinal Products for Human Use (2013/C 343/01) 202
Chapter 4 – Documentation 202
Chapter 10 – Specific Provisions for Brokers 204

9 UK Guidance on Brokering Medicines 206

Introduction 206
Brokering in Medicinal Products 206
Registration 207
Application for Brokering Registration 207
Criteria of Broker's Registration 208
Provision of Information 209
Good Distribution Practice 210
Management of Recall Activity by Brokers 210

Manufacture, Importation and Distribution of Active Substances 213

10 EU Legislation on Manufacture, Importation and Distribution of Active Substances 215

Directive 2001/83/EC, Title IV 215
Directive 2001/83/EC of the European Parliament and of the Council of 6 November 2001 on the Community code relating to medicinal products for human use 215
Title IV 216

11 UK Legislation on the Manufacture, Importation and Distribution of Active Substances 219

The Human Medicines Regulations 2012 (SI 2012/1916) 219

12 Guidelines of 19 March 2015 on Principles of Good Distribution Practice of Active Substances for Medicinal Products for Human Use (2015/C 95/01) 228

Introduction 228
Chapter 1 – Scope 229
Chapter 2 – Quality System 229
Chapter 3 – Personnel 230
Chapter 4 – Documentation 230
Chapter 5 – Premises and Equipment 232
Chapter 6 – Operations 232
Chapter 7 – Returns, Complaints and Recalls 235
Chapter 8 – Self-inspections 237
Annex 238

13 UK Guidance on the Manufacture, Importation and Distribution of Active Substances 240

Introduction 240
Registration 241
Conditions of Registration as a Manufacturer, Importer or Distributor of an Active Substance 244
GMP for Active Substances 245
GDP for Active Substances 246
Written Confirmation 246
Template for the 'written confirmation' for active substances exported to the European Union for medicinal products for human use, in accordance with Article 46b(2)(b) of Directive 2001/83/EC 247
Waiver from Written Confirmation 249
Procedure for Active Substance Importation 250
Procedure for Waiver from Written Confirmation 252
National Contingency Guidance 253
National Contingency Guidance Submission Template 255

Appendices 259

Appendix 1 261

Human and Veterinary Medicines Authorities in Europe 261

Appendix 2 267

List of Persons who can be Supplied with Medicines by Way of Wholesale Dealing (Human Medicines Regulations 2012) 267
All Medicines 267
Pharmacy Medicines 268
General Sale List Medicines 268
Other Relevant Provisions 268
Additional Provisions for Optometrists and Dispensing Opticians not included in Schedule 17 269
Schedule 17 Provisions 270

Appendix 3 287

Sources of Useful Information 287

Appendix 4 289

Licensing for Import into the UK and Export from the UK Including Introduced Medicine – Wholesale Supply Only 289

Appendix 5 293

Importation of Active Substances for Medicinal Products for Human Use – Questions and Answers 293

Appendix 6 302

Safety Features for Medicinal Products for Human Use – Questions and Answers 302

Index 317

EU referendum

Following the outcome of the EU referendum, the Medicines and Healthcare products Regulatory Agency (MHRA) is working closely with the Government to analyse the best options and opportunities available for the safe and effective regulation of medicines and medical devices in the UK. While negotiations continue, the UK remains a full and active member of the EU, with all the rights and obligations of EU membership firmly in place.

For further updates from the Medicines and Healthcare products Regulatory Agency concerning this matter please visit the website at www.gov.uk/mhra.

Preface to the 2017 edition

This is the 2017 edition of Rules and Guidance for Pharmaceutical Distributors (the "Green Guide"). This fourth edition has revised sections on:

- qualification of suppliers and customers;
- controls on certain medicinal products;
- parallel importation and parallel distribution;
- the application and inspection process for new licences: "what to expect";
- updated UK legislation; and
- temperature control and monitoring.

There are also new sections on:

- the guidelines on principles of Good Distribution Practice of active substances for medicinal products for human use (2015/C 95/01);
- data integrity;
- matters relating to unlicensed medicines;
- sourcing and exporting medicinal products – non-EEA countries; and
- the EU regulation on safety features for medicines.

Two Commission Q&As have been added:

- importation of active substances; and
- safety features for medicinal products.

There are also two new appendices on:

- sources of useful information; and
- licensing requirements for import into the UK and export from the UK including introduced medicine – wholesale supply only.

Changes to the Community Code

The Falsified Medicines Directive 2011/62/EU amends Directive 2001/83/EC in a number of places. The first tranche of these changes in relation to manufacturing, wholesale dealing, supervision and sanctions came into force from 2 January 2013 with others relating to the importation of active substances from countries outside the EEA taking effect from 2 July 2013. These provisions are implemented in the UK by Regulations amending the Human Medicines Regulations 2012.

The headline changes are as follows:

- The current regulatory expectation for the manufacturer of the medicinal product to have audited their suppliers of active substances for compliance with the relevant Good Manufacturing Practice (GMP) has been formalised, as is the requirement for the written confirmation of audit (the "QP Declaration", currently required as part of the marketing authorisation application). This audit may be undertaken by the manufacturer of the medicinal product, or by a suitable and appropriately experienced third party under contract to the manufacturer of the medicinal product.
- In addition a formal requirement for manufacturers of medicinal products (or a third party acting under contract) to audit their suppliers of active substances for compliance with the requirements of Good Distribution Practice (GDP) particular to active substances has also been introduced.
- Manufacturers, importers and distributors of active substances now have to be registered with the competent authority of the Member State in which they are established (in the UK this is MHRA). Registrations are entered onto a database operated by the European Medicines Agency, in a similar manner to the EudraGMDP database. The manufacturer of the medicinal product must verify that their suppliers of active substances are registered.
- Manufacturers of the medicinal product have to verify the authenticity and quality of the active substances and excipients they use.
- The manufacture of active substances for use in a licensed medicinal product must be in compliance with the relevant GMP. These standards are currently described in Part II of the EU Guidelines on GMP.
- Active substances imported from outside the EEA must have been manufactured in accordance with standards of GMP, at least equivalent to those in the EU, and from 2 July 2013 must be accompanied by a written confirmation that equivalent GMP standards and supervision apply in the exporting country, unless the active substance is sourced from a country, listed by the European Commission or exceptionally and where necessary; to ensure availability of medicinal products an EU GMP certificate for the site of manufacture is available.
- Manufacturers of medicinal products have to assess the risk to product quality presented by any excipients they use, by way of a formalised documented risk assessment, and ascertain the appropriate good manufacturing practices necessary to assure their safety and quality.

There is no explicit obligation for the medicinal product manufacturer to audit their suppliers of excipients, but the manufacturer is required to assure themselves that the appropriate good manufacturing practices are being applied.

- In support of the above changes the European Commission has adopted the following by means of delegated acts and guidelines:
  - Principles and guidelines for GMP for active substances.
  - GDP Guidelines for active substances.
  - Guidelines for the formal risk assessment process for excipients.
- Manufacturers, wholesale distributors and brokers of medicines have to inform the competent authority and the marketing authorisation holder (MAH) should information be obtained that products either manufactured under the scope of the manufacturing authorisation or received or offered may be falsified, whether those products are being distributed through the legitimate supply chain or by illegal means.
- Brokers of medicines have to be registered with the competent authority of the Member State in which they are established (in the UK this is MHRA) and must comply with the applicable aspects of GDP.
- Persons undertaking the wholesale distribution of medicinal products to third countries have to hold an authorisation and have to check that their customers are authorised to receive medicines. Where the medicinal products for export have been imported from a third country checks must also be made to ensure that the supplier is authorised to supply medicines.
- An extension of the requirement to notify MHRA and the MAH where a wholesale dealer imports from another EEA Member State into the UK a medicinal product that is the subject of a national marketing authorisation, and the importer is not the MAH or acting on the MAH's behalf, to centrally authorised products (those holding a marketing authorisation granted by the European Medicines Agency), and the introduction of the option for the competent authority to charge a fee for processing the notification. For products imported into the UK the competent authority would be either the European Medicines Agency or MHRA, depending on whether the product is centrally authorised or not.

Changes on the horizon

The requirements introduced by the Falsified Medicines Directive for medicinal products to bear safety features (set out below) are now the subject of a new separate Regulation from the European Commission. Regulation (EU) 2016/161 sets out the practicalities of the safety features. These safety features are a unique identifier (a 2D data matrix code and human readable information) which will be placed on medical products that can be scanned at fixed points along the supply chain and tamper evident features on the pack. The Regulation also sets out two lists: the list of prescription medicines that shall not bear the safety features and the list of non-prescription medicines that shall bear the safety features. The Delegated Regulation comes into force in the UK in February 2019.

Also under the Delegated Regulation:

- The Qualified Person is to ensure that the safety features have been affixed.
- Safety features are not to be removed or covered unless the manufacturing authorisation holder verifies that the medicinal product is authentic and has not been tampered with and that replacement safety features are equivalent and are applied in accordance with GMP.
- Wholesale dealers are to verify that any medicinal products they receive are not falsified, by checking that any "safety features" used on the outer packaging of a product are intact.
- Brokers and wholesale dealers are to record the batch numbers of, as a minimum, those products with safety features attached and for wholesale dealers to provide a record of batch numbers when supplying those products to their customers.

The Green Guide 2017

This is the fourth edition of the *Rules and Guidance for Pharmaceutical Distributors* known as the Green Guide issued by MHRA. In this edition we have reordered the sections to bring specific subject matter together to help the reader. As with the previous editions the Green Guide continues to bring together existing and revised Commission-written material concerning the distribution and brokering of human medicines and matters relating to the manufacture, importation and distribution of active substances.

Whilst UK medicines legislation bears directly on activities in the UK, users of the guide now more than ever need to be fully aware of the original obligations set out in EU medicines legislation that affect them in the current changing landscape. Therefore, the "Titles" or sections of Directive 2001/83/EC, as amended, dealing with wholesale distribution of products for human use, brokering finished medicinal products, and provisions relating to the manufacture, importation and distribution of active substances, remain in this edition.

In this edition the new guidelines on principles of Good Distribution Practice of active substances are included. These guidelines provide stand-alone guidance on Good Distribution Practice (GDP) for manufacturers, importers and distributors of active substances for medicinal products for human use. The EU Regulation on safety features for medicines is also partly reproduced prior to its transposition into UK medicines legislation by early 2019, together with the Commission's Q&As.

We have also incorporated new sections on some matters relating to unlicensed medicines, sourcing and exporting medicinal products (non-EEA countries), data integrity and two new appendices. One provides a source of useful information and the other provides a table on licensing requirements for import into the UK and export from the UK to complement the new section on sourcing and exporting (non-EEA countries). For importers of active substances we have included the Commission's Q&A on importation of active substances.

Updated sections have been provided for qualification of suppliers and customers, controls on certain medicinal products, parallel importation and parallel distribution, temperature control and monitoring of medicinal products and UK medicines legislation.

Although much of the text in this book is available in its original form in other places, including various websites, I remain pleased that the Green Guide continues to satisfy a demand for information in one authoritative and convenient place. The 2017 version will be available online, as part of "Medicines Complete" – a subscription-based database of leading medicines and healthcare references and in e-reader formats. I hope that this edition in its revised layout will continue to be useful.

Gerald Heddell
Director, Inspection, Enforcement and Standards Division
January 2017

Acknowledgements

To the European Commission for permission to reproduce the 1995–2017 text of the Directives (only European Union legislation printed in the paper edition of the Official Journal of the European Union is deemed authentic), Regulations, Guidelines and associated documents.

To the Heads of Medicines Agencies for permission to reproduce the names and addresses of other human and veterinary medicines authorities in Europe.

To Cogent for permission to reproduce the new Gold Standard role profile for the Responsible Person.

Feedback

Comments on the content or presentation of the Green Guide are encouraged and will be used to develop further editions. Your views are valued and both MHRA and Pharmaceutical Press would appreciate you taking the time to contact us. Please visit the feedback page at http://www.pharmpress.com/orangeguide-feedback or send your feedback to the address below:

"The Rules and Guidance for Pharmaceutical Distributors 2017"
Customer Services
MHRA
151 Buckingham Palace Road
Victoria
London SW1W 9SZ
UK
Tel.: +44 (0)20 3080 6000
Fax: +44 (0)20 3118 9803
Email: orange.guide@mhra.gsi.gov.uk

Introduction

The distribution network for medicinal products has become increasingly complex and now involves many different players.

The obligation on governments of all Member States of the European Union and European Economic Area (EEA)[1] to ensure that pharmaceutical wholesale distributors are authorised is stated in Title VII of the Directive 2001/83/EC. This Title requires all authorised wholesale distributors to have available a Responsible Person (RP) and to comply with the Commission guidelines on Good Distribution Practice (GDP).

Title VII Directive 2001/83/EC has been amended as a result of the Falsified Medicines Directive 2011/62/EU and now requires brokers of medicines within the Community to be registered with their competent authority and to comply with appropriate GDP requirements. Amended Title IV of the Directive 2001/83/EC also requires importers, manufacturers and distributors of active substances who are established in the Union to register their activity with the competent authority of the Member State in which they are established.

The Commission guidelines for GDP first issued in 1993 has been revised as a consequence of the Falsified Medicines Directive to assist wholesale distributors in conducting their activities and to prevent falsified medicines from entering the legal supply chain. The revised guidelines also provide specific rules for persons involved in activities in relation to the sale or purchase of medicinal products while not conducting a wholesale activity, i.e. brokers of medicines.

In the UK, the provisions for wholesale distributors, brokers of medicines and manufacturers, importers and distributors of active substances have been implemented by requirements and undertakings incorporated in the Human Medicines Regulations 2012 [SI 2012/1916]. Through these regulations compliance by wholesale distributors and brokers of medicines with the Commission guidelines of GDP is a statutory requirement.

---

[1] The member states of the European Community plus Iceland, Liechtenstein and Norway.

This publication brings together the new Commission guidelines on GDP, UK guidance on wholesale distribution practice and EU and UK legislation on wholesale distribution, which wholesale distributors are expected to follow when distributing medicinal products for human use. It is of particular relevance to authorised wholesale distributors and to their RPs, who have a responsibility for ensuring compliance with many of these regulatory requirements. Wholesale distributors are required to appoint an RP who has the knowledge and responsibility to ensure that correct procedures are followed during distribution. Updated notes on the qualifications and duties of RPs are included in this publication to assist this.

This publication is also of particular relevance to all brokers of medicines because it brings together UK guidance on brokering medicines and EU and UK legislation on brokering, which brokers are expected to follow when brokering medicinal products for human use.

It is also of relevance to importers and distributors of active substances. The manufacture of active substances for use in a licensed medicinal product must be in compliance with the relevant good manufacturing practice. These standards are currently described in Part II of the EU Commission Guidelines on Good Manufacturing Practice (GMP). Active substances imported from outside the EEA must have been manufactured in accordance with standards of GMP at least equivalent to those in the EU. From 2 July 2013 they must be accompanied by a written confirmation that equivalent GMP standards and supervision apply in the exporting country, unless the active substance is sourced from a country listed by the European Commission or exceptionally and where necessary to ensure availability of medicinal products an EU GMP certificate for the site of manufacture is available.

The European Commission has adopted principles and guidelines for GMP for active substances, GDP guidelines for active substances and guidelines for the formal risk assessment process for excipients by means of delegated acts and guidelines.

This guide brings together UK guidance and EU and UK legislation on the importation and distribution of active substances, which importers and distributors of active substances are expected to follow when supplying active substances for use in licensed medicinal products for human use. For completeness this guide also includes matters on what manufacturers of active substances have to comply with in order to provide further background information for importers and distributors of active substances.

The aim of GMP and GDP is to assure the quality of the medicinal product for the safety, well-being and protection of the patient. In achieving this aim it is impossible to over-emphasise the importance of individuals, at all levels, in the assurance of the quality of medicinal products. This is emphasised in the first principle in the EC Guide to GMP.

The great majority of reported defective medicinal products has resulted from human error or carelessness, not from failures in technology. All individuals involved with the distribution of medicinal products should bear this constantly in mind when performing their duties.

Glossary of Legislation

European Legislation

Council Directive 2001/83/EC on the Community code relating to medicinal products for human use

This legislation regulates the licensing of, manufacture of and wholesale dealing in medicinal products and registration, brokering of medicinal products and manufacture, importation and distribution of active substances within the European Community.

Council Directive 2003/94/EC laying down the principles and guidelines of good manufacturing practice in respect of medicinal products for human use and investigational medicinal products

This Directive lays down the principles and guidelines of good manufacturing practice in respect of medicinal products for human use, the manufacture of which requires an authorisation.

UK Legislation

The Human Medicines Regulations 2012 [SI 2012/1916]

The Regulations set out a comprehensive regime for the authorisation of medicinal products for human use, for the manufacture, import, distribution, sale and supply of those products, for their labelling and advertising, and for pharmacovigilance.

For the most part the Regulations implement Directive 2001/83/EC of the European Parliament and of the Council of 6 November 2001 on the community code relating to medicinal products for human use (as amended). They also provide for the enforcement in the United Kingdom of Regulation (EC) No 726/2004 laying down Community procedures for the authorisation and supervision of medicinal products for human and veterinary use and establishing a European Medicines Agency.

The Medicines (Products for Human Use) (Fees) Regulations 2016 [SI 2016/190]

These Regulations make provision for the fees payable under the Medicines Act 1971 and other fees payable in respect of EU obligations including those relating to authorisations, licences, certificates and registrations in respect of medicinal products for human use.

The Medicines for Human Use (Clinical Trials) Regulations 2004 [SI 2004/1031] as amended

These Regulations implement Directive 2001/20/EC on the approximation of laws, regulations and administrative provisions of the Member States relating to the implementation of good clinical practice in the conduct of clinical trials on medicinal products for human use.

The Unlicensed Medicinal Products for Human Use (Transmissible Spongiform Encephalopathies) (Safety) Regulations 2003 [SI 2003/1680]

Regulates the importation and marketing of unlicensed medicinal products for human use in order to minimise the risk of the transmission of transmissible spongiform encephalopathies via those products.

MHRA

CHAPTER

1

MHRA: Licensing, Inspection and Enforcement for Medicines

Contents

Overview of the Medicines and Healthcare products Regulatory Agency Group 3
Overview of MHRA 5
MHRA Innovation Office 6
MHRA supporting innovation in medicines 7
Inspection, Enforcement and Standards Division 7
Inspectorate 8
Good Manufacturing Practice (GMP) 8
Good Distribution Practice (GDP) 9
Good Laboratory Practice (GLP) 10
Good Clinical Practice (GCP) 10
Good Pharmacovigilance Practice (GPvP) 10
The Data Processing Group 10
Manufacturer's and wholesale dealer's licence/authorisations 11
Registrations 11
Export Certificates 12
Importing unlicensed medicines – import notifications 12
Defective Medicines Report Centre (DMRC) 12
Enforcement Group 13
Compliance Management and Inspection Action Group 14
Compliance Escalation process 14
Referral to Inspection Action Group 15
Advice 19

Overview of the Medicines and Healthcare products Regulatory Agency Group

The Medicines and Healthcare products Regulatory Agency regulates medicines, medical devices and blood components for transfusion in the UK.

Recognised globally as an authority in its field, the agency plays a leading role in protecting and improving public health and supports innovation through scientific research and development.

The agency has three centres:

- the Clinical Practice Research Datalink (CPRD), a data research service that aims to improve public health by using anonymised NHS clinical data;
- the National Institute for Biological Standards and Control (NIBSC), a global leader in the standardisation and control of biological medicines;
- the Medicines and Healthcare products Regulatory Agency (MHRA), the UK's regulator of medicines, medical devices and blood components for transfusion, responsible for ensuring their safety, quality and effectiveness.

The agency is responsible for:

- ensuring that medicines, medical devices and blood components for transfusion meet applicable standards of safety, quality and efficacy;
- ensuring that the supply chain for medicines, medical devices and blood components is safe and secure;
- promoting international standardisation and harmonisation to assure the effectiveness and safety of biological medicines;
- helping to educate the public and healthcare professionals about the risks and benefits of medicines, medical devices and blood components, leading to safer and more effective use;
- supporting innovation and research and development that's beneficial to public health; and
- influencing UK, EU and international regulatory frameworks so that they are risk proportionate and effective at protecting public health.

Medicines and Healthcare products Regulatory Agency

MHRA	National Institute for Biological Standards Board (NIBSC)	Clinical Practice Research Datalink (CPRD)
• Operating a system of licensing, classification, monitoring (post-marketing surveillance) and enforcement for medicines. • Discharging statutory obligations for medical devices, including designating and monitoring the performance of notified bodies. • Ensuring statutory compliance in medicines clinical trials and assessing medical device clinical trials proposals. • Promulgating good practice in the safe use of medicines and medical devices. • Regulating the safety and quality of blood and blood components. • Discharging the functions of the UK Good Laboratory Practice Monitoring Authority (GLPMA). • Managing the activities of the British Pharmacopoeia (BP).	• Devising and drawing up standards for the purity and potency of biological substances and designing appropriate test procedures. • Preparing, approving, holding and distributing standard preparations of biological substances. • Providing, or arranging for, the provision of laboratory testing facilities for the testing of biological substances, carrying out such testing, examining records of manufacture and quality control and reporting on the results. • Carrying out or arranging for the carrying out of research in connection with biological standards and control functions.	• Managing and designing CPRD services to maximise the way anonymised NHS clinical data can be linked to enables many types of observational research and deliver research outputs that are beneficial to improving and safeguarding public health.

Overview of MHRA

All licensed human medicines available in the UK are subject to rigorous scrutiny by MHRA before they can be used by patients. This ensures that human medicines meet acceptable standards on safety, quality and efficacy. It is the responsibility of MHRA and the expert advisory bodies set up by the Human Medicines Regulations 2012 [SI 2012/1916] (previously under the 1968 Medicines Act) to ensure that the sometimes difficult balance between safety and effectiveness is achieved. MHRA experts assess all applications for new human medicines to ensure they meet the required standards. This is followed up by a system of inspection and testing which continues throughout the lifetime of the medicine.

As the UK government's public health body which brings together the regulation of human medicines and medical devices, science and research, the roles of MHRA are to:

- license medicines, manufacturers and distributors;
- register brokers of finished medicines and manufacturers, importers and distributors of active substances;
- register websites that offer medicines for sale or supply to the public;
- regulate medical devices;
- approve UK clinical trials;
- monitor medicines and medical devices after licensing;
- ensure the safety and quality of blood;
- tackle illegal activity involving medicines, medical devices and blood;
- promote an understanding of the benefits and risks;
- facilitate the development of new medicines;
- support innovation in medicines and medical devices;
- be a leading provider of data and data services for healthcare research;
- work with international partners on issues; and
- provide a national voice for the benefits and risks of medicines, medical devices and medical technologies.

MHRA also hosts and supports a number of expert advisory bodies, including the Commission on Human Medicines (which replaced the Committee on the Safety of Medicines in 2005), and the British Pharmacopoeia Commission. In addition, as part of the European system of medicines approval, MHRA or other national bodies may be the Rapporteur or Co-rapporteur for any given pharmaceutical application, taking on the bulk of the verification work on behalf of all members, while the documents are still sent to other members as and where requested.

MHRA Innovation Office

One of the key objectives of MHRA is to support greater access to safe and effective medicinal products and medical devices. The timely introduction of innovative products benefits both patients and the public. This section provides information about MHRA's Innovation Office and highlights how you can get scientific and regulatory advice to support the development of innovational products.

MHRA's Innovation Office was set up by MHRA as part of the UK government's industrial strategy for life sciences and helps organisations that are developing innovative medicines, medical devices or using novel manufacturing processes to navigate the regulatory processes in order to be able to progress their products or technologies. Examples of innovative products include Advanced Therapy Medicinal Products (ATMPs), nano-technology, stratified medicines, novel drug/device combinations and advanced manufacturing.

If you are a pharmaceutical researcher, developer, manufacturer, etc. and you have a question about an innovative medicine, device or novel manufacturing process, contact the MHRA Innovation Office using one of the routes below.

After contacting the office you will receive a response within 20 working days. Depending on the nature of the query, your response will consist of either a simple answer or a recommended course of action, which may involve regulatory or scientific advice.

SCIENTIFIC AND REGULATORY ADVICE

MHRA offers regulatory and scientific advice. For medicines, we currently carry out over 250 scientific advice meetings a year. Joint meetings can be held with the National Institute for Health and Care Excellence (NICE) to consider health technology assessment issues at the same time.

SCIENTIFIC ADVICE FROM THE EU

The European Medicines Agency (EMA) offers a comprehensive scientific advice service, available to provide assistance during the initial development of a medicine and during the post-authorisation phase. The advice is provided by the EMA's Committee for Medicinal Products for Human Use (CHMP),[1] based on recommendations of the Scientific Advice Working Party (SAWP).[2] Further advice on accessing scientific advice can be obtained from the EMA.[3]

1 http://www.ema.europa.eu/ema/index.jsp?curl=pages/about_us/general/general_content_000094.jsp&mid=WC0b01ac0580028c79

2 http://www.ema.europa.eu/ema/index.jsp?curl=pages/contacts/CHMP/people_listing_000022.jsp&mid=WC0b01ac0580028d94

3 http://www.ema.europa.eu/ema/index.jsp?curl=pages/regulation/general/general_content_000049.jsp&mid=WC0b01ac05800229b9&jsenabled=true

SCIENTIFIC ADVICE FROM THE UK

MHRA offers a comprehensive scientific advice service which can assist companies in making decisions about a range of scientific and regulatory issues. Scientific advice can be requested at any stage of the initial development of a medicine and also during the pre-submission period. Please see MHRA webpages on "Scientific Advice".[4]

MHRA supporting innovation in medicines

MHRA supports innovation through:

Medicines

- membership on European committees (Commission on Human Medicinal Products (CHMP), CAT, COMP, PDCO) and CHMP Working Parties;[5]
- membership of European Good Distribution Practice (GDP)/Good Manufacturing Practice (GMP) Inspectors Working Group;
- advice given to the EMA Innovation Task Force on a wide range of innovative technologies and products;
- working within the ICH network;
- writing regulatory and scientific guidance documents via the ICH and CHMP; and
- membership of European GDP/GMP Inspectors Working Group.

Devices sector

- membership of European committees such as the Medical Devices Expert group (MDEG), Borderlines and Classification Group, New and Emerging Technologies Working Group (NET); and
- part of the International Medical Device Regulators Forum.

Inspection, Enforcement and Standards Division

MHRA's Inspection, Enforcement and Standards Division is responsible for ensuring compliance with the regulations and standards that apply to the manufacture, control and supply of medicines on the UK market.

[4] http://www.mhra.gov.uk/Howweregulate/Medicines/Licensingofmedicines/Informationforlicenceapplicants/Otherusefulservicesandinformation/Scientificadviceforlicenceapplicants/index.htm

[5] http://www.ema.europa.eu/ema/index.jsp?curl=pages/about_us/general/general_content_000217.jsp&mid=

Inspectorate

The Inspectorate Group in MHRA's Inspection, Enforcement and Standards Division comprises of dedicated units for Good Manufacturing Practice (GMP), Good Distribution Practice (GDP), Good Laboratory Practice (GLP), Good Clinical Practice (GCP) and Good Pharmacovigilance Practice (GPvP).

Good Manufacturing Practice (GMP)

GMP Inspectors conduct inspections of pharmaceutical manufacturers and other organisations to assess compliance with EC guidance on Good Manufacturing Practice (GMP) and the relevant details contained in marketing authorisations and Clinical Trials Authorisations. They ensure that medicines supplied in the UK and wider EU meet consistent high standards of quality, safety and efficacy. Overseas manufacturing sites to be named on UK or EU marketing authorisations are also required to pass an inspection prior to approval of the marketing authorisation application. Following approval, a risk-based inspection programme maintains on-going surveillance of UK and overseas manufacturing site compliance with EC GMP.

GMP Inspectors are responsible for inspecting and authorising a range of manufacturers of sterile and non-sterile dosage forms, biological products, investigational medicinal products, herbal products and active pharmaceutical ingredients, in addition to certain analytical laboratories. The manufacture of unlicensed medicines by holders of Manufacturer "Specials" Licences in the UK NHS and commercial sector is also inspected on a routine basis to assess compliance with relevant legislation and GMP.

The safety and quality of human blood for transfusion, or for further manufacture into blood-derived medicines, is ensured through inspections of relevant collection, processing, testing and storage activities at Blood Establishments and UK Hospital Blood Banks. These inspections assess compliance with specific UK and EU regulatory requirements, which take into account the detailed principles of GMP.

GMP Inspectors serve on a number of UK, EU and international technical and standards committees and provide help and advice to senior managers, Ministers and colleagues across the Agency, as necessary. Support and expertise is also provided to the inspection programmes of the European Medicines Agency (EMA), European Directorate for Quality of Medicines (EDQM) and the World Health Organization (WHO).

Good Distribution Practice (GDP)

GDP Inspectors conduct inspections of sites of wholesale dealers to assess compliance with EU Guidelines on Good Distribution Practice (GDP) and the conditions of a wholesale dealer's licence.

Inspectors will ensure that medicinal products are handled, stored and transported under conditions as prescribed by the marketing authorisation or product specification.

Inspections are undertaken of new applicants and then subsequently on a routine schedule based on a risk assessment of the site.

There are a number of developments that had an impact on GDP during 2013 and going forward including:

- the Human Medicines Regulations 2012 [SI 2012/1916] which came into force in August 2012, replacing the majority of the Medicines Act 1968 and its supporting legislation;
- the transposition of the Falsified Medicines Directive 2011/62/EU into UK medicines legislation which extends GDP to any person or entity who procures, stores or supplies medicinal products, for export to countries outside the EEA and to brokers of finished medicines within the EEA.
- the application of the revised EU Guidelines on GDP of 5 November 2013, which entered into force on 24 November 2013. This version replaced the earlier version, which entered into force on 8 September 2013, and introduced the following changes:
  - the maintenance of a quality system setting out responsibilities, processes and risk management principles in relation to wholesale activities;
  - suitable documentation which prevents errors from spoken communication;
  - sufficient competent personnel to carry out all the tasks for which the wholesale distributor is responsible;
  - adequate premises, installations and equipment so as to ensure proper storage and distribution of medicinal products;
  - appropriate management of complaints, returns, suspected falsified medicinal products and recalls;
  - outsourced activities correctly defined to avoid misunderstandings;
  - rules for transport in particular to protect medicinal products against breakage, adulteration and theft, and to ensure that temperature conditions are maintained within acceptable limits during transport;
  - specific rules for brokers (person involved in activities in relation to the sale or purchase of medicinal products, except for wholesale distribution, that do not include physical handling the products).

The revised EU Guidelines on GDP have been included in this publication.

Good Laboratory Practice (GLP)

GLP Inspectors conduct inspections of UK facilities that carry out non-clinical studies for submission to domestic and international regulatory authorities to assess the safety of new chemicals to humans, animals and the environment. These inspections are designed to assure that studies are performed in accordance with the relevant EC directives and the Organisation for Economic Co-operation and Development (OECD) principles as required by OECD Council acts relating to the Mutual Acceptance of Data. The range of test facilities to be monitored include those involved in the testing of human and veterinary pharmaceuticals, agrochemicals, food and feed additives, industrial chemicals and cosmetics.

Good Clinical Practice (GCP)

The GCP Inspectorate is responsible for inspecting clinical trials for compliance with Good Clinical Practice (GCP). Compliance with this good practice provides assurance that the rights, safety and well-being of trial subjects are protected, and that the results of the clinical trials are credible and accurate.

The function of the GCP Inspectorate is to assess the compliance of organisations with UK and EU legislation relating to the conduct of clinical trials in investigational medicinal products. This is achieved through carrying out inspections of sponsor organisations that hold Clinical Trial Authorisations (CTA), organisations that provide services to clinical trial sponsors or investigator sites.

Good Pharmacovigilance Practice (GPvP)

The Pharmacovigilance Inspectorate conducts inspections of the pharmacovigilance systems of marketing authorisation holders to assess compliance with the requirements of the UK and European legislation and guidelines relating to the monitoring of the safety of medicines given to patients.

The Data Processing Group

The Data Processing Group was formed in 2011 by consolidating the existing Process Licensing, Export Certificates, Import Notifications and Defective Medicines Reporting Centre functions to reside in a single Business Unit as part of the Inspection, Enforcement and Standards Division.

Manufacturer's and wholesale dealer's licence/authorisations

Manufacture of and wholesale dealing in medicinal products are licensable activities under UK and EU legislation. These licences are referred to as process licences and include a wide range of licences covering diverse activities listed below:

- Licences for the manufacture/importation of licensed medicinal products for human use, commonly abbreviated to MIA.
- "Specials" licences for the manufacture/importation of unlicensed medicinal products for human use, commonly abbreviated to MS.
- Authorisations for the manufacture/importation of Investigational Medicinal Products for human use, commonly abbreviated to MIA (IMP).
- Authorisations for the manufacture/importation of licensed medicinal products for veterinary use, commonly abbreviated to ManA.
- "Specials" licences for the manufacture of unlicensed medicinal products for veterinary use, commonly abbreviated to ManSA.
- Authorisation for the manufacture of Exempt Advanced Therapy Medicinal Products, commonly abbreviated to MeAT.
- Licences for the wholesale dealing of medicinal products for human use, commonly abbreviated to WDA(H) (including those covering unlicensed medicines obtained from another EEA Member State).
- Licences for the wholesale dealing/importation of medicinal products for veterinary use, commonly abbreviated to WDA(V).
- Blood Establishment Authorisations, commonly abbreviated to BEA.

The Data Processing team process all applications for new licences, variations to existing licences, changes of ownership, terminations, cancellations as well as suspensions and revocations on the instructions of the Inspection Action Group (IAG), making extensive use of computer technology to do so. They are also responsible for issuing Certificates of Good Manufacturing Practice (GMP) and Good Distribution Practice (GDP) on behalf of the GMP and GDP Inspectorate.

Registrations

The Falsified Medicines Directive 2011/62/EU made it a requirement that certain activities require registration and that a minimum of information be published on publically accessible registers. These activities include:

- brokering of finished human medicines;
- manufacture, importation and distribution of active substances;
- internet sales of medicines to members of the public.

The Data Processing team process all applications for registration, variations to existing registrations and annual compliance reports, terminations, cancellations, suspensions and revocations making extensive use of computer technology to do so.

Export Certificates

The Data Processing team are also responsible for issuing certificates in support of the World Health Organization (WHO) scheme on the quality of pharmaceutical products moving in international commerce (often referred to as export certificates):

- Certificate of a pharmaceutical product (CPP). This certificate complies with the WHO format.
- Certificate of licensing status (CLS). This certificate complies with the WHO format.
- Certificate of manufacturing status (CMS).
- Certificate for the importation of a pharmaceutical constituent (CPC).
- Statement of licensing status of a pharmaceutical product(s).

Importing unlicensed medicines – import notifications

Under regulation 46 of the Human Medicines Regulations 2012 [SI 2012/1916] a medicine must have a marketing authorisation (includes Product Licences) unless exempt. One of these exemptions, which is in regulation 167 to these regulations, is for the supply of unlicensed medicinal products for the special needs of individual patients, commonly, but incorrectly called "named patients". Prospective importers must hold a relevant licence and must notify MHRA of their intention to import:

- for import from within the European Economic Area (EEA), a Wholesale Dealer's Licence valid for import and handling unlicensed relevant medicinal products;
- for import from outside of the EEA, a Manufacturer's 'Specials' Licence valid for import.

The Data Processing team makes use of a bespoke computer system (INS) to enter the information, refer flagged requests to Pharmaceutical Assessors for assessment as required and issue confirmation letters authorising or rejecting importation.

Defective Medicines Report Centre (DMRC)

MHRA's Defective Medicines Report Centre (DMRC) plays a major part in the protection of public health by minimising the hazard to patients arising from the distribution of defective medicinal products. It does this by providing an emergency assessment and communications system between suppliers of medicinal products, the regulatory authorities and the users. It achieves this by receiving and assessing reports of suspected defective medicines, monitoring and as necessary advising and directing appropriate actions by the responsible authorisation holder and communicating the details of this action as necessary and with appropriate urgency to recipients of the products and other interested parties in the UK and elsewhere by means of drug alerts.

Manufacturers, importers and distributors are obliged to report to the licensing authority (MHRA) any quality defect in a medicinal product which could result in a recall or restriction on supply.

Where a defective medicine is considered to present a risk to public health, the marketing authorisation holder, or the manufacturer as appropriate, is responsible for recalling the affected batch(es) or, in extreme cases, removing all batches of the product from the market. The DMRC will normally support this action by the issue of a drug alert notification to healthcare professionals. Drug alerts are classed from 1 to 4 according to their criticality and the speed with which action must be taken to remove the defective medicine from the distribution chain and, where necessary, from the point of dispensing and use. This varies from immediate action for a Class 1 alert, to action within five days for a Class 3 alert. In some low-risk circumstances the product may be allowed to remain in the supply chain when the DMRC will issue a Class 4 "caution in use" alert.

The DMRC is also part of the European Rapid Alert System, and in the case of Class 1 and Class 2 will notify regulators in other countries using the European Rapid Alert System.

Enforcement Group

Medicines legislation contains statutory provisions to enforce the requirements of the Human Medicines Regulations 2012 [SI 2012/1916] and the remaining provisions of the 1968 Medicines Act.

This enforcement role is carried out by MHRA's Enforcement Group which comprises a Case Referrals Team, Intelligence Analysts, Investigations Team, Prosecution Unit and Policy/Relationships management.

The legislation confers certain powers, including rights of entry, powers of inspection, seizure, sampling and production of documents. Duly authorised Investigation Officers investigate cases using these powers and, where appropriate, criminal prosecutions are brought by the Crown Prosecution Service (CPS). MHRA investigators also investigate offences under other legislation such as the Fraud Act, Trademarks Act and the Offences Against the Person Act.

All reported breaches of medicines legislation are investigated. Reports are processed and risk assessed before a course of action is agreed in line with our published Enforcement Strategy.

The aim of the Intelligence Unit is to drive forward the implementation of intelligence-led enforcement and enable a more proactive approach to the acquisition and development of information. The Unit acts as a coordination point for all information-gathering activities and works in conjunction with a wide network of public and professional bodies and trade associations, e.g. UK Border Force, UK Border Agency, Department of Health, Trading Standards and Port Health Authorities, the Police Service; and professional organisations such as the General Pharmaceutical Council (GPhC), General Medical Council and the Association of the British Pharmaceutical Industry (ABPI). Additionally, there is a network of other regulatory agencies and law enforcement bodies within the European Community and in other countries through which the Enforcement Group can exchange information and follow trends in pharmaceutical crime.

The Enforcement Group monitors trends in pharmaceutical crime and coordinates initiatives to counteract criminal activity. In particular, the availability of counterfeit medicines is a key priority area and an anti-counterfeiting strategy has been agreed and implemented.

Compliance Management and Inspection Action Group

Compliance Escalation process

Compliance Escalation is a non-statutory process to take action in response to poor compliance which does not yet meet the threshold for consideration of adverse regulatory action. Compliance Escalation is managed via the Compliance Management Team (CMT), a non-statutory group of senior GMDP inspectors who coordinate and advise on compliance management activities arising from chronic or significant GMP deficiencies. The main aim of the Compliance Escalation process is to direct companies towards a state of compliance, thus avoiding the need for regulatory action and the potential adverse impact to patient health through lack of availability of medicines as a result of action against an authorisation, as well as avoiding reputational and commercial damage for the manufacturer and product owner.

The specific inspection case issues are considered by CMT, who make decisions in conjunction with the inspector regarding the proportionate inspection and non-inspection compliance management actions required. This may include making recommendations on close monitoring of compliance improvement work through inspection, requested meetings with company senior management, and correspondence with company senior management, alerting them to the compliance concerns, and clearly outlining the consequences of continued non-compliance.

Decisions on compliance management actions are communicated to the company, following consideration of any written responses to a post-inspection letter if relevant. The site inspector(s) and CMT will continue to monitor the effectiveness of these actions. The CMT process may also be initiated by the Inspection Action Group (IAG) following referral for significant or serious GMP deficiencies. In cases where consideration of adverse licensing action is no longer required due to improvements or mitigating actions, IAG may close their case referral and request that CMT maintain compliance management oversight until completion of the remediation plans. Upon satisfactory conclusion of the remediation work, the company will be returned to the routine risk-based inspection programme; however, referral for consideration of regulatory action may still occur if the required improvements are not achieved in a timely manner.

Referral to Inspection Action Group

Critical findings are routinely referred to MHRA's Inspection Action Group (IAG). The group considers referrals involving serious/critical GMP deficiencies. The primary objective of IAG is to protect public health by ensuring that licensable activities in the manufacturing and distribution of medicinal products meet the required regulatory standards.

The IAG is a non-statutory, multi-disciplinary group constituted to advise the Director of Inspection, Enforcement and Standards Division (IE&S) on any recommendation for regulatory or adverse licensing action appropriate to the Division. There are two distinct groups:

- IAG1 considers issues related to GMP, GDP and Blood Establishment Authorisations (GMP/GDP/BEA) and has 24 scheduled meetings per year, normally the first and third Tuesdays of each month; and
- IAG2 considers issues related to Good Clinical and Good Pharmacovigilance Practices (GCP/GPvP) and has 12 scheduled meetings per year, normally the fourth Tuesday of each month.

There may be occasions when there is not enough business to make a meeting worthwhile. Conversely, there have been several occasions where an urgent issue has arisen, and an emergency meeting called.

The following attend both IAG1 and IAG2:

- the regulatory unit manager or deputy (chair);
- at least one medical assessor from MHRA Licensing Division;
- at least one pharmaceutical assessor from MHRA Licensing Division;
- a solicitor from Government Legal Services;
- at least one member of the IAG Secretariat;

- any inspector making a referral to the IAG;
- a representative from Enforcement Operations (if required); and
- members of MHRA staff may also attend for training purposes.

In addition, the following will attend IAG1:

- an expert/senior GMP inspector;
- an expert/senior GDP inspector;
- if required a representative from the Veterinary Medicines Directorate (external link); and
- an expert in blood and blood products (if required).

In addition, the following will attend IAG2:

- an expert/senior GCP inspector;
- an expert/senior GPvP inspector;
- the Clinical Trials Unit manager or deputy; and
- the Pharmacovigilance Risk Management Group manager or deputy.

REASONS FOR REFERRAL

This will usually happen if, during the inspection process, the inspector has identified one or more critical "deficiencies" as a result of a Good Practice Standards (GxP) inspection. However, a referral may also be made as a result of a licence variation, the failure to contact an organisation, the refusal of an organisation to accept an inspection, the outcome of enforcement activity, the outcome of a product recall or from an issue raised by another Member State.

The company will be informed during the closing meeting of an inspection that a referral to IAG will be made. This will be further confirmed in the post-inspection letter. From that point, correspondence between the company and the Agency should go via the IAG Secretariat.

How the process works:

(1) Inspection reveals serious (critical) deficiencies and informs company of IAG referral.
(2) Post-inspection letter is issued to company signed by responsible operations manager/lead inspector.
(3) IAG discusses the case at its next available meeting (if necessary an emergency meeting can be called).
(4) IAG proposes its action to the divisional director for approval.
(5) The referred company is informed of IAG action and next steps.
(6) Actions are followed up at subsequent meetings until the situation is resolved.
(7) The matter is kept on the agenda until IAG is satisfied that the referral can be closed.

IAG POSSIBLE ACTIONS

IAG1 (GMP/GDP/BEA)

- refusal to grant a licence or a variation;
- proposal to suspend the licence for a stated period or vary the licence;
- notification of immediate suspension of the licence for a stated period (no longer than three months);
- proposal to revoke the licence;
- notice of suspension, variation or revocation;
- action to remove a Qualified Person/Responsible Person (QP/RP) from the licence;
- issue a Cease and Desist order in relation to a Blood Establishment Authorisation;
- issue a warning letter to the company/individual;
- request a written justification for actions of a QP/RP;
- referral of a QP to his/her professional body;
- increased inspection frequency;
- statement of non-compliance with GMP or GDP;
- restricted GMP and GDP certificates;
- request the company/individual attend a meeting at the Agency; or
- refer to the Enforcement Group for further consideration.

IAG2 (GCP/GPvP)

- issue an infringement notice in relation to a clinical trial;
- suspend or revoke a clinical trial authorisation;
- further follow-up inspections, or triggered inspections at related organisations (e.g. issues in GCP may trigger a GMP inspection);
- referral to CHMP (Committee for Medicinal Products for Human Use) for consideration for or against a marketing authorisation (e.g. suspended, varied or revoked);
- liaison and coordinated action with EMA (European Medicines Agency) and other Member States regarding concerns;
- refer the case to the EMA for consideration of the use of the EU Infringement Regulation (which could result in a fine);
- request a written justification for action of a QPPV (Qualified Person responsible for pharmacovigilance);
- request the company/individual attend a meeting at the Agency; or
- refer to the Enforcement Group for further consideration.

In the case of inspections in third countries:

- a refusal to name a site on a marketing authorisation;
- a recommendation that a site be removed from a marketing authorisation;
- the issuing of a GMP non-compliance statement;

- in the case of an adverse (voluntary or triggered) active pharmaceutical ingredients (API) inspection, this could result in the removal of the API site from the marketing authorisation; or
- in the case of an adverse (voluntary or triggered) investigational medicinal products (IMP) inspection, this could result in the suspension of a clinical trial.

In all cases, an action could result in the withdrawal of product (API, IMP, etc.) from the market. This specific action is, however, handled by the Defective Medicines Reporting Centre (DMRC) rather than directly by IAG.

LEGAL BASIS FOR THIS ACTION

The legal basis for licensing action is contained in:

- The Human Medicines Regulations 2012 [SI 2012/1916];
- The Medicines for Human Use (Clinical Trials) Regulations 2004 [SI 2004/1031]; and
- The Blood Safety and Quality Regulations 2005 [SI 2005/50].

WHAT A LICENCE HOLDER SHOULD DO IF REFERRED TO IAG

In the first instance, a referral should be treated as a requirement to immediately correct the deficiencies identified during the inspection and report completed actions to the IAG Secretariat/Inspectorate as soon as possible.

If the referral results in an immediate suspension of a manufacturing/ wholesale dealer's licence, there are no rights of appeal for the immediate suspension (which can last no longer than three months). During this time a company should be focused on correcting the inspection deficiencies.

If the referral results in a proposed suspension, variation or revocation of a licence a company will have the following appeal options prior to a decision being made:

- may make written representations to the licensing authority (MHRA); or
- may appear before and be heard by a person appointed for the purpose by the licensing authority (a fee of £10,000 will be charged for a person-appointed request).

If a company submits written representations, the licensing authority shall take those representations into account before determining the matter. In practice, this means that any proposed action will not be progressed until the written representations have been reviewed and considered by IAG and a recommendation made to the Divisional Director on whether to proceed with the action or not.

If a company submits a request for a Person Appointed Hearing, this will be taken forward by the Panel Secretariat, which sits within MHRA's

Policy Division. Any proposed action will not be progressed until the Person Appointed Hearing has taken place.

It should be noted that a Person Appointed Hearing will only offer its opinion into whether a licence condition has been contravened. A final decision on whether to suspend or revoke a licence will still rest with the licensing authority, who will take the report of the Person Appointed Hearing into account.

Follow-up actions that may be taken:

- a re-inspection to ensure corrective actions implemented;
- request for regular updates on the corrective action plan;
- the issue of a short dated GMP certificate;
- recommended increase of inspection frequency;
- continued monitoring of the company by IAG via inspectorate updates; or
- if serious and persistent non-compliance continues referral for consideration of criminal prosecution.

Contact for further information: IAGSecretariat@mhra.gsi.gov.uk

Advice

MHRA publishes a series of Guidance Notes relating to its statutory functions. Those of particular interest to manufacturers and wholesale dealers include:

GN 5	Notes for applicants and holders of a manufacturer's licence
GN 6	Notes for applicants and holders of a wholesale dealer's licence (WDA(H)) or broker registration
GN 8	A guide to what is a medicinal product
GN 14	The supply of unlicensed medicinal products "Specials"

These Guidance Notes and a list of others available may be obtained from MHRA's website or from MHRA's Customer Services Team.

Contact details are as follows:

Address:
Customer Services, MHRA, 151 Buckingham Palace Road, Victoria, London SW1W 9SZ, UK
Telephone: +44 (0)20 3080 6000 (weekdays 09:00–17:00)
Fax: +44 (0)20 3118 9803
E-mail: info@mhra.gsi.gov.uk
Website: www.mhra.gov.uk

Wholesale Distribution of Medicines

CHAPTER

2

EU Legislation on Wholesale Distribution

Contents

Article 20 41
Article 21 41
Article 22 42
Article 23 42
Article 24 43
Chapter VI: Modalities of Verification of the Safety Features and Decommissioning of the Unique Identifier by Persons Authorised or Entitled to Supply Medicinal Products to the Public 43
Article 25 43
Article 26 44
Article 27 45
Article 28 45
Article 29 45
Article 30 46
Chapter VII: Establishment, Management and Accessibility of the Repositories System 46
Article 31 46
Article 32 47
Article 33 47
Article 34 49
Article 35 50
Article 36 52
Article 37 53
Article 38 54
Article 39 54
Chapter VIII: Obligations of Marketing Authorisation Holders, Parallel Importers and Parallel Distributors 55
Article 40 55
Article 41 55
Article 42 55
Chapter IX: Obligations of the National Competent Authorities 56
Article 43 56
Article 44 56
Chapter X: Lists of Derogations and Notifications to the Commission 57
Article 45 57
Article 46 57
Article 47 58
Chapter XI: Transitional Measures and Entry Into Force 58
Article 48 58
Article 49 58
Article 50 59
Annex I 60
Annex II 61
Annex III 62
Annex IV 63

DIRECTIVE 2001/83/EC, AS AMENDED, TITLE VII, WHOLESALE DISTRIBUTION AND BROKERING MEDICINES

Directive 2001/83/EC of the European Parliament and of the Council of 6 November 2001 on the Community code relating to medicinal products for human use as amended

Editor's note Title VII of this Directive is reproduced below. Title VII has been amended by Directive 2011/62/EU and Directive 2012/26/EU. A new article 85a extends the need for a Wholesale Dealer's Licence for the export of medicine to a non-EEA country and for "introduced" medicines that are sourced from a non-EEA country for export back to a non-EEA country. A new article 85b introduces measures for persons brokering medicinal products. Reference should be made to the full Directive 2001/83/EC as amended for the preamble, definitions, and general and final provisions.

Title VII: Wholesale Distribution and Brokering of Medicinal Products

Article 76

1 Without prejudice to Article 6, Member States shall take all appropriate action to ensure that only medicinal products in respect of which a marketing authorization has been granted in accordance with Community law are distributed on their territory.

2 In the case of wholesale distribution and storage, medicinal products shall be covered by a marketing authorisation granted pursuant to Regulation (EC) No. 726/2004 or by the competent authorities of a Member State in accordance with this Directive.

3 Any distributor, not being the marketing authorisation holder, who imports a medicinal product from another Member State shall notify the marketing authorisation holder and the competent authority in the Member State to which the medicinal product will be imported of his intention to import that product. In the case of medicinal products which have not been granted an authorisation pursuant to Regulation (EC) No 726/2004, the notification to the competent authority shall be without prejudice to additional procedures provided for in the legislation of that

Member State and to fees payable to the competent authority for examining the notification.

4 In the case of medicinal products which have been granted an authorisation pursuant to Regulation (EC) No 726/2004, the distributor shall submit the notification in accordance with paragraph 3 of this Article to the marketing authorisation holder and the Agency. A fee shall be payable to the Agency for checking that the conditions laid down in Union legislation on medicinal products and in the marketing authorisations are observed.

Article 77

1 Member States shall take all appropriate measures to ensure that the wholesale distribution of medicinal products is subject to the possession of an authorisation to engage in activity as a wholesaler in medicinal products, stating the premises located on their territory for which it is valid.

2 Where persons authorized or entitled to supply medicinal products to the public may also, under national law, engage in wholesale business, such persons shall be subject to the authorization provided for in paragraph 1.

3 Possession of a manufacturing authorization shall include authorization to distribute by wholesale the medicinal products covered by that authorization. Possession of an authorization to engage in activity as a wholesaler in medicinal products shall not give dispensation from the obligation to possess a manufacturing authorization and to comply with the conditions set out in that respect, even where the manufacturing or import business is secondary.

4 Member States shall enter the information relating to the authorisations referred to in paragraph 1 of this Article in the Union database referred to in Article 111(6). At the request of the Commission or any Member State, Member States shall provide all appropriate information concerning the individual authorisations which they have granted under paragraph 1 of this Article.

5 Checks on the persons authorised to engage in activity as a wholesaler in medicinal products, and the inspection of their premises, shall be carried out under the responsibility of the Member State which granted the authorisation for premises located on its territory.

6 The Member State which granted the authorization referred to in paragraph 1 shall suspend or revoke that authorization if the conditions of authorization cease to be met. It shall forthwith inform the other Member States and the Commission thereof.

7 Should a Member State consider that, in respect of a person holding an authorization granted by another Member State under the terms of paragraph 1, the conditions of authorization are not, or are no longer met, it shall forthwith inform the Commission and the other Member State involved. The latter shall take the measures necessary and shall inform the Commission and the first Member State of the decisions taken and the reasons for those decisions.

Article 78

Member States shall ensure that the time taken for the procedure for examining the application for the distribution authorization does not exceed 90 days from the day on which the competent authority of the Member State concerned receives the application.

The competent authority may, if need be, require the applicant to supply all necessary information concerning the conditions of authorization. Where the authority exercises this option, the period laid down in the first paragraph shall be suspended until the requisite additional data have been supplied.

Article 79

In order to obtain the distribution authorization, applicants must fulfil the following minimum requirements:

(a) they must have suitable and adequate premises, installations and equipment, so as to ensure proper conservation and distribution of the medicinal products;
(b) they must have staff, and in particular, a qualified person designated as responsible, meeting the conditions provided for by the legislation of the Member State concerned;
(c) they must undertake to fulfil the obligations incumbent on them under the terms of Article 80.

Article 80

Holders of the distribution authorization must fulfil the following minimum requirements:

(a) they must make the premises, installations and equipment referred to in Article 79(a) accessible at all times to the persons responsible for inspecting them;

(**b**) they must obtain their supplies of medicinal products only from persons who are themselves in possession of the distribution authorization or who are exempt from obtaining such authorization under the terms of Article 77(3);

(**c**) they must supply medicinal products only to persons who are themselves in possession of the distribution authorization or who are authorized or entitled to supply medicinal products to the public in the Member State concerned;

(**ca**) they must verify that the medicinal products received are not falsified by checking the safety features on the outer packaging, in accordance with the requirements laid down in the delegated acts referred to in Article 54a(2);

(**d**) they must have an emergency plan which ensures effective implementation of any recall from the market ordered by the competent authorities or carried out in cooperation with the manufacturer or marketing authorization holder for the medicinal product concerned;

(**e**) they must keep records either in the form of purchase/sales invoices or on computer, or in any other form, giving for any transaction in medicinal products received, dispatched or brokered at least the following information:

- date,
- name of the medicinal product,
- quantity received, supplied or brokered,
- name and address of the supplier or consignee, as appropriate,
- batch number of the medicinal products at least for products bearing the safety features referred to in point (o) of Article 54;

(**f**) they must keep the records referred to under (e) available to the competent authorities, for inspection purposes, for a period of five years;

(**g**) they must comply with the principles and guidelines of good distribution practice for medicinal products as laid down in Article 84.

(**h**) they must maintain a quality system setting out responsibilities, processes and risk management measures in relation to their activities;

(**i**) they must immediately inform the competent authority and, where applicable, the marketing authorisation holder, of medicinal products they receive or are offered which they identify as falsified or suspect to be falsified.

For the purposes of point (b), where the medicinal product is obtained from another wholesale distributor, wholesale distribution authorisation holders must verify compliance with the principles and guidelines of good distribution practices by the supplying wholesale distributor. This includes verifying whether the supplying wholesale distributor holds a wholesale distribution authorisation.

Where the medicinal product is obtained from the manufacturer or importer, wholesale distribution authorisation holders must verify that the manufacturer or importer holds a manufacturing authorisation.

Where the medicinal product is obtained through brokering, the wholesale distribution authorisation holders must verify that the broker involved fulfils the requirements set out in this Directive.

Article 81

With regard to the supply of medicinal products to pharmacists and persons authorised or entitled to supply medicinal products to the public, Member States shall not impose upon the holder of a distribution authorisation which has been granted by another Member State any obligation, in particular public service obligations, more stringent than those they impose on persons whom they have themselves authorised to engage in equivalent activities.

The holder of a marketing authorisation for a medicinal product and the distributors of the said medicinal product actually placed on the market in a Member State shall, within the limits of their responsibilities, ensure appropriate and continued supplies of that medicinal product to pharmacies and persons authorised to supply medicinal products so that the needs of patients in the Member State in question are covered.

The arrangements for implementing this Article should, moreover, be justified on grounds of public health protection and be proportionate in relation to the objective of such protection, in compliance with the Treaty rules, particularly those concerning the free movement of goods and competition.

Article 82

For all supplies of medicinal products to a person authorized or entitled to supply medicinal products to the public in the Member State concerned, the authorized wholesaler must enclose a document that makes it possible to ascertain:

- batch number of the medicinal products at least for products bearing the safety features referred to in point (o) of Article 54;
- the date;
- the name and pharmaceutical form of the medicinal product;
- the quantity supplied;
- the name and address of the supplier and consignor.

Member States shall take all appropriate measures to ensure that persons authorized or entitled to supply medicinal products to the public are able to provide information that makes it possible to trace the distribution path of every medicinal product.

Article 83

The provisions of this Title shall not prevent the application of more stringent requirements laid down by Member States in respect of the wholesale distribution of:

- narcotic or psychotropic substances within their territory;
- medicinal products derived from blood;
- immunological medicinal products;
- radiopharmaceuticals.

Article 84

The Commission shall publish guidelines on good distribution practice. To this end, it shall consult the Committee for Medicinal Products for Human Use and the Pharmaceutical Committee established by Council Decision 75/320/EEC.[1]

Article 85

This Title shall apply to homeopathic medicinal products.

Article 85a

In the case of wholesale distribution of medicinal products to third countries, Article 76 and point (c) of the first paragraph of Article 80 shall not apply. Moreover, points (b) and (ca) of the first paragraph of Article 80 shall not apply where a product is directly received from a third country but not imported. However, in that case wholesale distributors shall ensure that the medicinal products are obtained only from persons who are authorised or entitled to supply medicinal products in accordance with the applicable legal and administrative provisions of the third country concerned. Where wholesale distributors supply medicinal products to persons in third countries, they shall ensure that such supplies are only made to persons who are authorised or entitled to receive medicinal products for wholesale distribution or supply to the public in accordance with the applicable legal and administrative provisions of the third country concerned. The requirements set out in Article 82 shall apply to the supply

[1] OJ L 147, 9.6.1975, p. 23.

of medicinal products to persons in third countries authorised or entitled to supply medicinal products to the public.

Article 85b

1 Persons brokering medicinal products shall ensure that the brokered medicinal products are covered by a marketing authorisation granted pursuant to Regulation (EC) No 726/2004 or by the competent authorities of a Member State in accordance with this Directive.

Persons brokering medicinal products shall have a permanent address and contact details in the Union, so as to ensure accurate identification, location, communication and supervision of their activities by competent authorities.

The requirements set out in points (d) to (i) of Article 80 shall apply *mutatis mutandis* to the brokering of medicinal products.

2 Persons may only broker medicinal products if they are registered with the competent authority of the Member State of their permanent address referred to in paragraph 1. Those persons shall submit, at least, their name, corporate name and permanent address in order to register. They shall notify the competent authority of any changes thereof without unnecessary delay.

Persons brokering medicinal products who had commenced their activity before 2 January 2013 shall register with the competent authority by 2 March 2013.

The competent authority shall enter the information referred to in the first subparagraph in a register that shall be publicly accessible.

3 The guidelines referred to in Article 84 shall include specific provisions for brokering.

4 This Article shall be without prejudice to Article 111. Inspections referred to in Article 111 shall be carried out under the responsibility of the Member State where the person brokering medicinal products is registered.

If a person brokering medicinal products does not comply with the requirements set out in this Article, the competent authority may decide to remove that person from the register referred to in paragraph 2. The competent authority shall notify that person thereof.

COMMISSION DELEGATED REGULATION (EU) 2016/161 OF 2 OCTOBER 2015 SUPPLEMENTING DIRECTIVE 2001/83/EC OF THE EUROPEAN PARLIAMENT AND OF THE COUNCIL BY LAYING DOWN DETAILED RULES FOR THE SAFETY FEATURES APPEARING ON THE PACKAGING OF MEDICINAL PRODUCTS FOR HUMAN USE

Editor's note

Safety Features Legislation

The Falsified Medicines Directive 2011/62/EU (FMD) which amends Directive 2001/83/EC introduces obligatory 'safety features' to allow verification of the authenticity of medicinal products. FMD stipulates that:

(1) medicinal products subject to prescription shall bear the safety features, including the unique identifier, unless they have been listed by the Commission in a delegated act;

(2) medicinal products not subject to prescription shall not bear the safety features, unless they have been listed by the Commission in a delegated act.

In accordance with Article 54a(2) of Directive 2001/83/EC, the Commission has adopted a Delegated Regulation (EU) 2016/161 which sets out the practicalities of the safety features.

These safety features are:

- a unique identifier (a 2D data matrix code and human readable information) which will be placed on medical products that can be scanned at fixed points along the supply chain
- tamper evident features on the pack

The unique identifier comprises:

- a product code which allows the identification of at least the name of the medicine, the common name, the pharmaceutical form, the strength, the pack size, and the pack type
- a serial number which is a numeric or alphanumeric sequence of a maximum of 20 characters randomly generated
- a batch number
- an expiry date

The Delegated Regulation also includes two lists:

- The list of prescription medicines that shall not bear the safety features and
- The list of non-prescription medicines that shall bear the safety features

The Delegated Regulation comes into force in the UK in 2019.

An extract of the Delegated Regulation (EU) 2016/161 is provided below.

Reference should be made to the full Regulation for the preamble, and final provisions.

The Commission question and answer document to implement the Regulation is provided within Appendix 6 and can also be found at:

http://ec.europa.eu/health/files/falsified_medicines/qa_safetyfeature_v5_0.pdf

CHAPTER I: SUBJECT MATTER AND DEFINITIONS

Article 1

SUBJECT MATTER

This Regulation lays down:

(a) the characteristics and technical specifications of the unique identifier that enables the authenticity of medicinal products to be verified and individual packs to be identified;
(b) the modalities for the verification of the safety features;
(c) the provisions on the establishment, management and accessibility of the repositories system where the information on the safety features shall be contained;
(d) the list of medicinal products and product categories subject to prescription which shall not bear the safety features;
(e) the list of medicinal products and product categories not subject to prescription which shall bear the safety features;
(f) the procedures for the notification to the Commission by national competent authorities of non-prescription medicinal products judged at risk of falsification and prescription medicinal products not deemed at risk of falsification in accordance with the criteria set out in Article 54a(2)(b) of Directive 2001/83/EC;
(g) the procedures for a rapid evaluation of and decision on the notifications referred to in point (f) of this Article.

Article 2

SCOPE

(1) This Regulation applies to:
  (a) medicinal products subject to prescription which shall bear safety features on their packaging pursuant to Article 54a(1) of Directive 2001/83/EC, unless included in the list set out in Annex I to this Regulation;

(b) medicinal products not subject to prescription included in the list set out in Annex II to this Regulation;
(c) medicinal products to which Member States have extended the scope of application of the unique identifier or of the anti-tampering device in accordance with Article 54a(5) of Directive 2001/83/EC.

(2) For the purposes of this Regulation, where reference is made to the packaging in a provision of this Regulation, the provision shall apply to outer packaging or to the immediate packaging if the medicinal product has no outer packaging.

Article 3

DEFINITIONS

(1) For the purposes of this Regulation, the definitions in Article 1 of Directive 2001/83/EC shall apply.

(2) The following definitions shall apply:
(a) 'unique identifier' means the safety feature enabling the verification of the authenticity and the identification of an individual pack of a medicinal product;
(b) 'anti-tampering device' means the safety feature allowing the verification of whether the packaging of a medicinal product has been tampered with;
(c) 'decommissioning of a unique identifier' means the operation changing the active status of a unique identifier stored in the repositories system referred to in Article 31 of this Regulation to a status impeding any further successful verification of the authenticity of that unique identifier;
(d) 'active unique identifier' means a unique identifier which has not been decommissioned or which is no longer decommissioned;
(e) 'active status' means the status of an active unique identifier stored in the repositories system referred to in Article 31;
(f) 'healthcare institution' means a hospital, in- or outpatient clinic or health centre.

CHAPTER II: TECHNICAL SPECIFICATIONS OF THE UNIQUE IDENTIFIER

Article 4

COMPOSITION OF THE UNIQUE IDENTIFIER

The manufacturer shall place on the packaging of a medicinal product a unique identifier which complies with the following technical specifications:

(a) The unique identifier shall be a sequence of numeric or alphanumeric characters that is unique to a given pack of a medicinal product.
(b) The unique identifier shall consist of the following data elements:
   (i) a code allowing the identification of at least the name, the common name, the pharmaceutical form, the strength, the pack size and the pack type of the medicinal product bearing the unique identifier ('product code');
   (ii) a numeric or alphanumeric sequence of maximum 20 characters, generated by a deterministic or a non-deterministic randomisation algorithm ('serial number');
   (iii) a national reimbursement number or other national number identifying the medicinal product, if required by the Member State where the product is intended to be placed on the market;
   (iv) the batch number;
   (v) the expiry date.
(c) The probability that the serial number can be guessed shall be negligible and in any case lower than one in ten thousand.
(d) The character sequence resulting from the combination of the product code and the serial number shall be unique to a given pack of a medicinal product until at least one year after the expiry date of the pack or five years after the pack has been released for sale or distribution in accordance with Article 51(3) of Directive 2001/83/EC, whichever is the longer period.
(e) Where the national reimbursement number or other national number identifying the medicinal product is contained in the product code, it is not required to be repeated within the unique identifier.

Article 5

CARRIER OF THE UNIQUE IDENTIFIER

(1) Manufacturers shall encode the unique identifier in a two-dimensional barcode.
(2) The barcode shall be a machine-readable Data Matrix and have error detection and correction equivalent to or higher than those of the Data Matrix ECC200. Barcodes conforming to the International Organization for Standardisation/International Electrotechnical Commission standard ('ISO/IEC') 16022:2006 shall be presumed to fulfil the requirements set out in this paragraph.
(3) Manufacturers shall print the barcode on the packaging on a smooth, uniform, low-reflecting surface.
(4) When encoded in a Data Matrix, the structure of the unique identifier shall follow an internationally-recognised, standardised data syntax

and semantics ('coding scheme') which allows the identification and accurate decoding of each data element of which the unique identifier is composed, using common scanning equipment. The coding scheme shall include data identifiers or application identifiers or other character sequences identifying the beginning and the end of the sequence of each individual data element of the unique identifier and defining the information contained in those data elements. Unique identifiers having a coding scheme conforming to ISO/IEC 15418:2009 shall be presumed to fulfil the requirements set out in this paragraph.

(5) When encoded in a Data Matrix as data element of a unique identifier, the product code shall follow a coding scheme and begin with characters specific to the coding scheme used. It shall also contain characters or character sequences identifying the product as a medicinal product. The resulting code shall be less than 50 characters and be globally unique. Product codes which conform to the ISO/IEC 15459-3:2014 and ISO/IEC 15459-4:2014 shall be presumed to fulfil the requirements set out in this paragraph.

(6) Where necessary, different coding schemes may be used within the same unique identifier provided that the decoding of the unique identifier is not hindered. In that case, the unique identifier shall contain standardised characters permitting the identification of the beginning and the end of the unique identifier as well as the beginning and the end of each coding scheme. Where containing multiple coding schemes, unique identifiers which conform to ISO/IEC 15434:2006 shall be presumed to fulfil the requirements set out in this paragraph.

Article 6

QUALITY OF THE PRINTING OF THE TWO-DIMENSIONAL BARCODE

(1) Manufacturers shall evaluate the quality of the printing of the Data Matrix by assessing at least the following Data Matrix parameters:
(a) the contrast between the light and dark parts;
(b) the uniformity of the reflectance of the light and dark parts;
(c) the axial non-uniformity;
(d) the grid non-uniformity;
(e) the unused error correction;
(f) the fixed pattern damage;
(g) the capacity of the reference decode algorithm to decode the Data Matrix.

(2) Manufacturers shall identify the minimum quality of the printing which ensures the accurate readability of the Data Matrix throughout the supply chain until at least one year after the expiry date of the pack

or five years after the pack has been released for sale or distribution in accordance with Article 51(3) of Directive 2001/83/EC, whichever is the longer period.

(3) When printing the Data Matrix, manufacturers shall not use a quality of the printing lower than the minimum quality referred to in paragraph 2.

(4) A quality of printing rated at least 1,5 in accordance with ISO/IEC 15415:2011 shall be presumed to fulfil the requirements set out in this Article.

Article 7

HUMAN-READABLE FORMAT

(1) Manufacturers shall print the following data elements of the unique identifier on the packaging in human-readable format:
   (a) the product code;
   (b) the serial number;
   (c) the national reimbursement number or other national number identifying the medicinal product, if required by the Member State where the product is intended to be placed on the market and not printed elsewhere on the packaging.

(2) Paragraph 1 shall not apply where the sum of the two longest dimensions of the packaging equals or is less than 10 centimetres.

(3) Where the dimensions of the packaging allow it, the human-readable data elements shall be adjacent to the two-dimensional barcode carrying the unique identifier.

Article 8

ADDITIONAL INFORMATION IN THE TWO-DIMENSIONAL BARCODE

Manufacturers may include information other than the unique identifier in the two-dimensional barcode carrying the unique identifier, where permitted by the competent authority in accordance with Title V of Directive 2001/83/EC.

Article 9

BARCODES ON THE PACKAGING

Medicinal products having to bear the safety features pursuant to Article 54a of Directive 2001/83/EC shall not bear on their packaging, for the purpose of their identification and verification of their authenticity, any other visible two-dimensional barcode than the two-dimensional barcode carrying the unique identifier.

CHAPTER III: GENERAL PROVISIONS ON THE VERIFICATION OF THE SAFETY FEATURES

Article 10

VERIFICATION OF THE SAFETY FEATURES

When verifying the safety features, manufacturers, wholesalers and persons authorised or entitled to supply medicinal products to the public shall verify the following:

(a) the authenticity of the unique identifier;
(b) the integrity of the anti-tampering device.

Article 11

VERIFICATION OF THE AUTHENTICITY OF THE UNIQUE IDENTIFIER

When verifying the authenticity of a unique identifier, manufacturers, wholesalers and persons authorised or entitled to supply medicinal products to the public shall check the unique identifier against the unique identifiers stored in the repositories system referred to in Article 31. A unique identifier shall be considered authentic when the repositories system contains an active unique identifier with the product code and serial number that are identical to those of the unique identifier being verified.

Article 12

UNIQUE IDENTIFIERS WHICH HAVE BEEN DECOMMISSIONED

A medicinal product bearing a unique identifier which has been decommissioned shall not be further distributed or supplied to the public except in any of the following situations:

(a) the unique identifier was decommissioned in accordance with Article 22(a) and the medicinal product is distributed for the purpose of exporting it outside the Union;
(b) the unique identifier was decommissioned earlier than the time of supplying the medicinal product to the public, pursuant to Articles 23, 26, 28 or 41;
(c) the unique identifier was decommissioned in accordance with Article 22(b) or (c) or Article 40, and the medicinal product is provided to the person responsible for its disposal;

(d) the unique identifier was decommissioned in accordance with Article 22(d) and the medicinal product is provided to the national competent authorities.

Article 13

REVERSING THE STATUS OF A DECOMMISSIONED UNIQUE IDENTIFIER

(1) Manufacturers, wholesalers and persons authorised or entitled to supply medicinal products to the public may only revert the status of a decommissioned unique identifier to an active status if the following conditions are fulfilled:
   (a) the person performing the reverting operation is covered by the same authorisation or entitlement and operates in the same premises as the person that decommissioned the unique identifier;
   (b) the reverting of the status takes place not more than 10 days after the unique identifier was decommissioned;
   (c) the pack of medicinal product has not expired;
   (d) the pack of medicinal product has not been registered in the repositories system as recalled, withdrawn, intended for destruction or stolen and the person performing the reverting operation does not have knowledge that the pack is stolen;
   (e) the medicinal product has not been supplied to the public.

(2) Medicinal products bearing a unique identifier which cannot be reverted to an active status because the conditions set out in paragraph 1 are not fulfilled shall not be returned to saleable stock.

CHAPTER IV: MODALITIES OF VERIFICATION OF THE SAFETY FEATURES AND DECOMMISSIONING OF THE UNIQUE IDENTIFIER BY MANUFACTURERS

Article 14

VERIFICATION OF THE TWO-DIMENSIONAL BARCODE

The manufacturer placing the safety features shall verify that the two-dimensional barcode carrying the unique identifier complies with Articles 5 and 6, is readable and contains the correct information.

Article 15

RECORD KEEPING

The manufacturer placing the safety features shall keep records of every operation he performs with or on the unique identifier on a pack of medicinal product for at least one year after the expiry date of the pack or five years after the pack has been released for sale or distribution in accordance with Article 51(3) of Directive 2001/83/EC, whichever is the longer period, and shall provide those records to competent authorities on request.

Article 16

VERIFICATIONS TO BE PERFORMED BEFORE REMOVING OR REPLACING THE SAFETY FEATURES

(1) Before removing or covering, either fully or partially, the safety features in accordance with Article 47a of Directive 2001/83/EC, the manufacturer shall verify the following:
  (a) the integrity of the anti-tampering device;
  (b) the authenticity of the unique identifier and decommission it if replaced.

(2) Manufacturers holding both a manufacturing authorisation according to Article 40 of Directive 2001/83/EC and an authorisation to manufacture or import investigational medicinal products to the Union as referred to in Article 61 of Regulation (EU) No 536/2014 of the European Parliament and of the Council[(3)] shall verify the safety features and decommission the unique identifier on a pack of medicinal product before repackaging or re-labelling it for the purpose of using it as authorised investigational medicinal product or authorised auxiliary medicinal product.

Article 17

EQUIVALENT UNIQUE IDENTIFIER

When placing an equivalent unique identifier for the purposes of complying with Article 47a(1)(b) of Directive 2001/83/EC, the manufacturer shall verify that the structure and composition of the unique identifier placed on the packaging complies, with regard to the product code and the national reimbursement number or other national number identifying the medicinal product, with the requirements of the Member State where the medicinal product is intended to be placed on the market, so that that unique identifier can be verified for authenticity and decommissioned.

Article 18

ACTIONS TO BE TAKEN BY MANUFACTURERS IN CASE OF TAMPERING OR SUSPECTED FALSIFICATION

Where a manufacturer has reason to believe that the packaging of the medicinal product has been tampered with, or the verification of the safety features shows that the product may not be authentic, the manufacturer shall not release the product for sale or distribution and shall immediately inform the relevant competent authorities.

Article 19

PROVISIONS APPLICABLE TO A MANUFACTURER DISTRIBUTING HIS PRODUCTS BY WHOLESALE

Where a manufacturer distributes his products by wholesale, Article 20(a), and Articles 22, 23 and 24 shall apply to him in addition to Articles 14 to 18.

CHAPTER V: MODALITIES OF VERIFICATION OF THE SAFETY FEATURES AND DECOMMISSIONING OF THE UNIQUE IDENTIFIER BY WHOLESALERS

Article 20

VERIFICATION OF THE AUTHENTICITY OF THE UNIQUE IDENTIFIER BY WHOLESALERS

A wholesaler shall verify the authenticity of the unique identifier of at least the following medicinal products in his physical possession:

(a) medicinal products returned to him by persons authorised or entitled to supply medicinal products to the public or by another wholesaler;
(b) medicinal products he receives from a wholesaler who is neither the manufacturer nor the wholesaler holding the marketing authorisation nor a wholesaler who is designated by the marketing authorisation holder, by means of a written contract, to store and distribute the products covered by his marketing authorisation on his behalf.

Article 21

DEROGATIONS FROM ARTICLE 20(B)

Verification of the authenticity of the unique identifier of a medicinal product is not required under Article 20(b) in any of the following situations:

(a) that medicinal product changes ownership but remains in the physical possession of the same wholesaler;
(b) that medicinal product is distributed within the territory of a Member State between two warehouses belonging to the same wholesaler or the same legal entity, and no sale takes place.

Article 22

DECOMMISSIONING OF UNIQUE IDENTIFIERS BY WHOLESALERS

A wholesaler shall verify the authenticity of and decommission the unique identifier of the following medicinal products:

(a) products which he intends to distribute outside of the Union;
(b) products which have been returned to him by persons authorised or entitled to supply medicinal products to the public or another wholesaler and cannot be returned to saleable stock;
(c) products which are intended for destruction;
(d) products which, while in his physical possession, are requested as a sample by competent authorities;
(e) products which he intends to distribute to the persons or institutions referred to in Article 23, where required by national legislation in accordance with the same Article.

Article 23

PROVISIONS TO ACCOMMODATE SPECIFIC CHARACTERISTICS OF MEMBER STATES' SUPPLY CHAINS

Member States may require, where necessary to accommodate the particular characteristics of the supply chain on their territory, that a wholesaler verifies the safety features and decommissions the unique identifier of a medicinal product before he supplies that medicinal product to any of the following persons or institutions:

(a) persons authorised or entitled to supply medicinal products to the public who do not operate within a healthcare institution or within a pharmacy;
(b) veterinarians and retailers of veterinary medicinal products;
(c) dental practitioners;
(d) optometrists and opticians;
(e) paramedics and emergency medical practitioners;
(f) armed forces, police and other governmental institutions maintaining stocks of medicinal products for the purposes of civil protection and disaster control;

(g) universities and other higher education establishments using medicinal products for the purposes of research and education, with the exceptions of healthcare institutions;
(h) prisons;
(i) schools;
(j) hospices;
(k) nursing homes.

Article 24

ACTIONS TO BE TAKEN BY WHOLESALERS IN CASE OF TAMPERING OR SUSPECTED FALSIFICATION

A wholesaler shall not supply or export a medicinal product where he has reason to believe that its packaging has been tampered with, or where the verification of the safety features of the medicinal product indicates that the product may not be authentic. He shall immediately inform the relevant competent authorities.

CHAPTER VI: MODALITIES OF VERIFICATION OF THE SAFETY FEATURES AND DECOMMISSIONING OF THE UNIQUE IDENTIFIER BY PERSONS AUTHORISED OR ENTITLED TO SUPPLY MEDICINAL PRODUCTS TO THE PUBLIC

Article 25

OBLIGATIONS OF PERSONS AUTHORISED OR ENTITLED TO SUPPLY MEDICINAL PRODUCTS TO THE PUBLIC

(1) Persons authorised or entitled to supply medicinal products to the public shall verify the safety features and decommission the unique identifier of any medicinal product bearing the safety features they supply to the public at the time of supplying it to the public.
(2) Notwithstanding paragraph 1, persons authorised or entitled to supply medicinal products to the public operating within a healthcare institution may carry out that verification and decommissioning at any time the medicinal product is in the physical possession of the healthcare institution, provided that no sale of the medicinal product takes place between the delivery of the product to the healthcare institution and the supplying of it to the public.
(3) In order to verify the authenticity of the unique identifier of a medicinal product and decommission that unique identifier, persons authorised or entitled to supply medicinal products to the public shall connect to the repositories system referred to in Article 31 through the

national or supranational repository serving the territory of the Member State in which they are authorised or entitled.

(4) They shall also verify the safety features and decommission the unique identifier of the following medicinal products bearing the safety features:
   (a) medicinal products in their physical possession that cannot be returned to wholesalers or manufacturers;
   (b) medicinal products that, while in their physical possession, are requested as samples by competent authorities, in accordance with national legislation;
   (c) medicinal products which they supply for subsequent use as authorised investigational medicinal products or authorised auxiliary medicinal products as defined in Articles 2(2)(9) and (10) of Regulation (EU) No 536/2014.

Article 26

DEROGATIONS FROM ARTICLE 25

(1) Persons authorised or entitled to supply medicinal products to the public are exempted from the obligation to verify the safety features and decommission the unique identifier of medicinal products provided to them as free samples in accordance with Article 96 of Directive 2001/83/EC.

(2) Persons authorised or entitled to supply medicinal products to the public who do not operate within a healthcare institution or within a pharmacy are exempted from the obligation to verify the safety features and decommission the unique identifier of medicinal products where that obligation has been placed on wholesalers by national legislation in accordance with Article 23.

(3) Notwithstanding Article 25, Member States may decide, where necessary to accommodate the particular characteristics of the supply chain on their territory, to exempt a person authorised or entitled to supply medicinal products to the public operating within a healthcare institution from the obligations of verification and decommissioning of the unique identifier, provided that the following conditions are met:
   (a) the person authorised or entitled to supply medicinal products to the public obtains the medicinal product bearing the unique identifier through a wholesaler belonging to the same legal entity as the healthcare institution;

(b) the verification and decommissioning of the unique identifier is performed by the wholesaler that supplies the product to the healthcare institution;
(c) no sale of the medicinal product takes place between the wholesaler supplying the product and that healthcare institution;
(d) the medicinal product is supplied to the public within that healthcare institution.

Article 27

OBLIGATIONS WHEN APPLYING THE DEROGATIONS

Where the verification of the authenticity and decommissioning of the unique identifier is carried out earlier than referred to in Article 25(1), pursuant to Articles 23 or 26, the integrity of the anti-tampering device shall be verified at the time the medicinal product is supplied to the public.

Article 28

OBLIGATIONS WHEN SUPPLYING ONLY PART OF A PACK

Notwithstanding Article 25(1), where persons authorised or entitled to supply medicinal products to the public supply only part of a pack of a medicinal product the unique identifier of which is not decommissioned, they shall verify the safety features and decommission that unique identifier when the pack is opened for the first time.

Article 29

OBLIGATIONS IN CASE OF INABILITY TO VERIFY THE AUTHENTICITY AND DECOMMISSION THE UNIQUE IDENTIFIER

Notwithstanding Article 25(1), where technical problems prevent persons authorised or entitled to supply medicinal products to the public from verifying the authenticity of and decommissioning a unique identifier at the time the medicinal product bearing that unique identifier is supplied to the public, those persons authorised or entitled to supply medicinal products to the public shall record the unique identifier and, as soon as the technical problems are solved, verify the authenticity of and decommission the unique identifier.

Article 30

ACTIONS TO BE TAKEN BY PERSONS AUTHORISED OR ENTITLED TO SUPPLY MEDICINAL PRODUCTS TO THE PUBLIC IN CASE OF SUSPECTED FALSIFICATION

Where persons authorised or entitled to supply medicinal products to the public have reason to believe that the packaging of the medicinal product has been tampered with, or the verification of the safety features of the medicinal product indicates that the product may not be authentic, those persons authorised or entitled to supply medicinal products to the public shall not supply the product and shall immediately inform the relevant competent authorities.

CHAPTER VII: ESTABLISHMENT, MANAGEMENT AND ACCESSIBILITY OF THE REPOSITORIES SYSTEM

Article 31

ESTABLISHMENT OF THE REPOSITORIES SYSTEM

(1) The repositories system where the information on the safety features shall be contained, pursuant to Article 54a(2)(e) of Directive 2001/83/EC, shall be set up and managed by a non-profit legal entity or non-profit legal entities established in the Union by manufacturers and marketing authorisation holders of medicinal products bearing the safety features.

(2) In setting up the repositories system, the legal entity or entities referred to in paragraph 1 shall consult at least wholesalers, persons authorised or entitled to supply medicinal products to the public and relevant national competent authorities.

(3) Wholesalers and persons authorised or entitled to supply medicinal products to the public are entitled to participate in the legal entity or entities referred to in paragraph 1, on a voluntary basis, at no cost.

(4) The legal entity or entities referred to on paragraph 1 shall not require manufacturers, marketing authorisation holders, wholesalers or persons authorised or entitled to supply medicinal products to the public to be members of a specific organisation or organisations in order to use the repository system.

(5) The costs of the repositories system shall be borne by the manufacturers of medicinal products bearing the safety features, in accordance with Article 54a(2)(e) of Directive 2001/83/EC.

Article 32

STRUCTURE OF THE REPOSITORIES SYSTEM

(1) The repositories system shall be composed of the following electronic repositories:
  (a) a central information and data router ('hub');
  (b) repositories which serve the territory of one Member State ('national repositories') or the territory of multiple Member States ('supranational repositories'). Those repositories shall be connected to the hub.

(2) The number of national and supranational repositories shall be sufficient to ensure that the territory of every Member State is served by one national or supranational repository.

(3) The repositories system shall comprise the necessary information technology infrastructure, hardware and software to enable the execution of the following tasks:
  (a) upload, collate, process, modify and store the information on the safety features that enables the verification of the authenticity and identification of medicinal products;
  (b) identify an individual pack of a medicinal product bearing the safety features and verify the authenticity of the unique identifier on that pack and decommission it at any point of the legal supply chain.

(4) The repositories system shall include the application programming interfaces allowing wholesalers or persons authorised or entitled to supply medicinal products to the public to query the repositories system by means of software, for the purposes of verifying the authenticity of the unique identifiers and of decommissioning them in the repositories system. The application programming interfaces shall also allow national competent authorities to access the repositories system by means of software, in accordance with Article 39.

The repositories system shall also include graphical user interfaces providing direct access to the repositories system in accordance with Article 35(1)(i).

The repositories system shall not include the physical scanning equipment used for reading the unique identifier.

Article 33

UPLOADING OF INFORMATION IN THE REPOSITORIES SYSTEM

(1) The marketing authorisation holder or, in case of parallel imported or parallel distributed medicinal products bearing an equivalent unique identifier for the purposes of complying with Article 47a of Directive

2001/83/EC, the person responsible for placing those medicinal products on the market, shall ensure that the information referred to in paragraph 2 is uploaded to the repositories system before the medicinal product is released for sale or distribution by the manufacturer, and that it is kept up to date thereafter.

The information shall be stored in all national or supranational repositories serving the territory of the Member State or Member States where the medicinal product bearing the unique identifier is intended to be placed on the market. The information referred to in paragraphs 2(a) to (d) of this Article, with the exception of the serial number, shall also be stored in the hub.

(2) For a medicinal product bearing a unique identifier, at least the following information shall be uploaded to the repositories system:
   (**a**) the data elements of the unique identifier in accordance with Article 4(b);
   (**b**) the coding scheme of the product code;
   (**c**) the name and the common name of the medicinal product, the pharmaceutical form, the strength, the pack type and the pack size of the medicinal product, in accordance with the terminology referred to in Article 25(1)(b) and (e) to (g) of the Commission Implementing Regulation (EU) No 520/2012 [(1)];
   (**d**) the Member State or Member States where the medicinal product is intended to be placed on the market;
   (**e**) where applicable, the code identifying the entry corresponding to the medicinal product bearing the unique identifier in the database referred to in Article 57(1)(l) of Regulation (EC) No 726/2004 of the European Parliament and the Council [(2)];
   (**f**) the name and address of the manufacturer placing the safety features;
   (**g**) the name and address of the marketing authorisation holder;
   (**h**) a list of wholesalers who are designated by the marketing authorisation holder, by means of a written contract, to store and distribute the products covered by his marketing authorisation on his behalf.

---

(1) Commission Implementing Regulation (EU) No 520/2012 of 19 June 2012 on the performance of pharmacovigilance activities provided for in Regulation (EC) No 726/2004 of the European Parliament and of the Council and Directive 2001/83/EC of the European Parliament and of the Council (OJ L 159, 20.6.2012, p. 5).

(2) Regulation (EC) No 726/2004 of the European Parliament and of the Council of 31 March 2004 laying down Community procedures for the authorisation and supervision of medicinal products for human and veterinary use and establishing a European Medicines Agency (OJ L 136, 30.4.2004, p. 1).

(3) The information referred to in paragraph 2 shall be uploaded to the repositories system either through the hub or through a national or supranational repository.

Where the upload is performed through the hub, the hub shall store a copy of the information referred to in paragraph 2(a) to (d), with the exception of the serial number, and transfer the complete information to all national or supranational repositories serving the territory of the Member State or Member States where the medicinal product bearing the unique identifier is intended to be placed on the market.

Where the upload is performed through a national or supranational repository, that repository shall immediately transfer to the hub a copy of the information referred to in paragraph 2(a) to (d), with the exception of the serial number, using the data format and data exchange specifications defined by the hub.

(4) The information referred to in paragraph 2 shall be stored in the repositories where it was originally uploaded for at least one year after the expiry date of the medicinal product or five years after the product has been released for sale or distribution in accordance with Article 51(3) of Directive 2001/83/EC, whichever is the longer period.

Article 34

FUNCTIONING OF THE HUB

(1) Each national or supranational repository composing the repositories system shall exchange data with the hub using the data format and data exchange modalities defined by the hub.

(2) When the authenticity of the unique identifier cannot be verified because a national or supranational repository does not contain a unique identifier with the product code and serial number that are identical to those of the unique identifier being verified, the national or supranational repository shall transfer the query to the hub in order to verify whether that unique identifier is stored elsewhere in the repositories system.

When the hub receives the query, the hub shall identify, on the basis of the information contained therein, all national or supranational repositories serving the territory of the Member State or Member States where the medicinal product bearing the unique identifier was intended to be placed on the market, and shall transfer the query to those repositories.

The hub shall subsequently transfer the reply of those repositories to the repository which initiated the query.

(3) Where notified by a national or supranational repository of the change of status of a unique identifier, the hub shall ensure the synchronisation of that status between those national or supranational repositories serving the territory of the Member State or Member States where the medicinal product bearing the unique identifier was intended to be placed on the market.

(4) When it receives the information referred to in Article 35(4), the hub shall ensure the electronic linking of the batch numbers before and after the repackaging or re-labelling operations with the set of unique identifiers decommissioned and with the set of equivalent unique identifiers placed.

Article 35

CHARACTERISTICS OF THE REPOSITORIES SYSTEM

(1) Each repository in the repositories system shall satisfy all of the following conditions:

(a) it shall be physically located in the Union;

(b) it shall be set up and managed by a non-profit legal entity established in the Union by manufacturers and marketing authorisation holders of medicinal products bearing the safety features and, where they have chosen to participate, wholesalers and persons authorised or entitled to supply medicinal products to the public;

(c) it shall be fully interoperable with the other repositories composing the repositories system; for the purposes of this Chapter, interoperability means the full functional integration of, and electronic data exchange between repositories regardless of the service provider used;

(d) it shall allow the reliable electronic identification and authentication of individual packs of medicinal products by manufacturers, wholesalers and persons authorised or entitled to supply medicinal products to the public, in accordance with the requirements of this Regulation;

(e) it shall have application programming interfaces able to transfer and exchange data with the software used by wholesalers, persons authorised or entitled to supply medicinal products to the public and, where applicable, national competent authorities;

(f) when wholesalers and persons authorised or entitled to supply medicinal products to the public query the repository for the purposes of verification of authenticity and decommissioning of a unique identifier, the response time of the repository, not considering the speed of the internet connection, shall be lower than 300 milliseconds in at least 95% of queries. The repository performance shall allow wholesalers and persons authorised or entitled to supply medicinal products to the public to operate without significant delay;

(g) it shall maintain a complete record ('audit trail') of all operations concerning a unique identifier, of the users performing those operations and the nature of the operations; the audit trail shall be created when the unique identifier is uploaded to the repository and be maintained until at least one year after the expiry date of the medicinal product bearing the unique identifier or five years after the product has been released for sale or distribution in accordance with Article 51(3) of Directive 2001/83/EC, whichever is the longer period;

(h) in accordance with Article 38, its structure shall be such as to guarantee the protection of personal data and information of a commercially confidential nature and the ownership and confidentiality of the data generated when manufacturers, marketing authorisation holders, wholesalers and persons authorised or entitled to supply medicinal products to the public interact with it;

(i) it shall include graphical user interfaces providing direct access to it to the following users verified in accordance with Article 37(b):

   (i) wholesalers and persons authorised or entitled to supply medicinal products to the public, for the purposes of verifying the authenticity of the unique identifier and decommissioning it in case of failure of their own software;

   (ii) national competent authorities, for the purposes referred to in Article 39;

(2) Where the status of a unique identifier on a medicinal product intended to be placed on the market in more than one Member State changes in a national or supranational repository, that repository shall immediately notify the change of status to the hub, except in case of decommissioning by marketing authorisation holders in accordance with Articles 40 or 41.

(3) National or supranational repositories shall not allow the upload or storage of a unique identifier containing the same product code and serial number as another unique identifier already stored therein.

(4) For each batch of repackaged or relabelled packs of a medicinal product on which equivalent unique identifiers were placed for the purposes of complying with Article 47a of Directive 2001/83/EC, the person responsible for placing the medicinal product on the market shall inform the hub of the batch number or numbers of the packs which are to be repackaged or relabelled and of the unique identifiers on those packs. He shall additionally inform the hub of the batch

number of the batch resulting from the repackaging or re-labelling operations and the equivalent unique identifiers in that batch.

Article 36

OPERATIONS OF THE REPOSITORIES SYSTEM

The repositories system shall provide for at least the following operations:

(a) the repeated verification of the authenticity of an active unique identifier in accordance with Article 11;
(b) the triggering of an alert in the system and in the terminal where the verification of the authenticity of a unique identifier is taking place when such verification fails to confirm that the unique identifier is authentic in accordance with Article 11. Such an event shall be flagged in the system as a potential incident of falsification except where the product is indicated in the system as recalled, withdrawn or intended for destruction;
(c) the decommissioning of a unique identifier in accordance with the requirements of this Regulation;
(d) the combined operations of identification of a pack of a medicinal product bearing a unique identifier and verification of the authenticity and decommissioning of that unique identifier;
(e) the identification of a pack of a medicinal product bearing a unique identifier and the verification of the authenticity and the decommissioning of that unique identifier in a Member State which is not the Member State where the medicinal product bearing that unique identifier was placed on the market;
(f) the reading of the information contained in the two-dimensional barcode encoding the unique identifier, the identification of the medicinal product carrying the barcode and the verification of the status of the unique identifier, without triggering the alert referred to in point (b) of this Article;
(g) without prejudice to Article 35(1)(h), the access by verified wholesalers to the list of wholesalers referred to in Article 33(2)(h) for the purposes of determining whether they have to verify the unique identifier of a given medicinal product.
(h) the verification of the authenticity of a unique identifier and its decommissioning by manually querying the system with the data elements of the unique identifier;
(i) the immediate provision of information concerning a given unique identifier to the national competent authorities and the European Medicines Agency, upon request;

(j) the creation of reports that allow competent authorities to verify compliance of individual marketing authorisation holders, manufacturers, wholesalers and persons authorised or entitled to supply medicinal products to the public with the requirements of this Regulation or to investigate potential incidents of falsification;
(k) the reverting of the status of a unique identifier from decommissioned to active, subject to the conditions referred to in Article 13;
(l) the indication that a unique identifier has been decommissioned;
(m) the indication that a medicinal product has been recalled, withdrawn, stolen, exported, requested as a sample by national competent authorities, indicated as a free sample by the marketing authorisation holder, or is intended for destruction;
(n) the linking, by batches of medicinal products, of the information on unique identifiers removed or covered to the information on the equivalent unique identifiers placed on those medicinal products for the purposes of complying with Article 47a of Directive 2001/83/EC.
(o) the synchronisation of the status of a unique identifier between the national or supranational repositories serving the territory of the Member States where that medicinal product is intended to be placed on the market.

Article 37

OBLIGATIONS OF LEGAL ENTITIES ESTABLISHING AND MANAGING A REPOSITORY WHICH IS PART OF THE REPOSITORIES SYSTEM

Any legal entity establishing and managing a repository which is part of the repositories system shall perform the following actions:

(a) inform the relevant national competent authorities of its intention to physically locate the repository or part of it in their territory and notify them once the repository becomes operational;
(b) put in place security procedures ensuring that only users whose identity, role and legitimacy has been verified can access the repository or upload the information referred to in Article 33(2);
(c) continuously monitor the repository for events alerting to potential incidents of falsification in accordance to Article 36(b);
(d) provide for the immediate investigation of all potential incidents of falsification flagged in the system in accordance with Article 36(b) and for the alerting of national competent authorities, the European Medicines Agency and the Commission should the falsification be confirmed;

(e) carry out regular audits of the repository to verify compliance with the requirements of this Regulation. Audits shall take place at least annually for the first five years after this Regulation becomes applicable in the Member State where the repository is physically located, and at least every three years thereafter. The outcome of those audits shall be provided to competent authorities upon request;
(f) make the audit trail referred to in Article 35(1)(g) immediately available to competent authorities upon their request;
(g) make the reports referred to in Article 36(j) available to competent authorities upon their request.

Article 38

DATA PROTECTION AND DATA OWNERSHIP

(1) Manufacturers, marketing authorisation holders, wholesalers and persons authorised or entitled to supply medicinal products to the public shall be responsible for any data generated when they interact with the repositories system and stored in the audit trail. They shall only have ownership of and access to those data, with the exception of the information referred to in Article 33(2) and the information on the status of a unique identifier.
(2) The legal entity managing the repository where the audit trail is stored shall not access the audit trail and the data contained therein without the written agreement of the legitimate data owners except for the purpose of investigating potential incidents of falsification flagged in the system in accordance with Article 36(b).

Article 39

ACCESS BY NATIONAL COMPETENT AUTHORITIES

A legal entity establishing and managing a repository used to verify the authenticity of or decommission the unique identifiers of medicinal products placed on the market in a Member State shall grant access to that repository and to the information contained therein, to competent authorities of that Member State for the following purposes:

(a) supervising the functioning of the repositories and investigating potential incidents of falsification;
(b) reimbursement;
(c) pharmacovigilance or pharmacoepidemiology.

CHAPTER VIII: OBLIGATIONS OF MARKETING AUTHORISATION HOLDERS, PARALLEL IMPORTERS AND PARALLEL DISTRIBUTORS

Article 40

PRODUCTS RECALLED, WITHDRAWN OR STOLEN

The marketing authorisation holder or, in case of parallel imported or parallel distributed medicinal products bearing an equivalent unique identifier for the purposes of complying with Article 47a of Directive 2001/83/EC, the person responsible for placing those medicinal products on the market shall promptly take all the following measures:

(a) ensure the decommissioning of the unique identifier of a medicinal product which is to be recalled or withdrawn, in every national or supranational repository serving the territory of the Member State or Member States in which the recall or the withdrawal is to take place;
(b) ensure the decommissioning of the unique identifier, where known, of a medicinal product which has been stolen, in every national or supranational repository in which information on that product is stored;
(c) indicate in the repositories referred to in points (a) and (b) that that product has been recalled or withdrawn or stolen, where applicable.

Article 41

PRODUCTS TO BE SUPPLIED AS FREE SAMPLES

The marketing authorisation holder intending to supply any of his medicinal products as a free sample in accordance with Article 96 of Directive 2001/83/EC shall, where that product bears the safety features, indicate it as a free sample in the repositories system and ensure the decommissioning of its unique identifier before providing it to the persons qualified to prescribe it.

Article 42

REMOVAL OF UNIQUE IDENTIFIERS FROM THE REPOSITORIES SYSTEM

The marketing authorisation holder or, in case of parallel imported or parallel distributed medicinal products bearing an equivalent unique identifier for the purposes of complying with Article 47a of Directive 2001/83/EC, the person responsible for placing those medicinal products on the market shall not upload unique identifiers in the repositories system before having removed from therein, where present, older unique identifiers containing the same product code and serial number as the unique identifiers being uploaded.

CHAPTER IX: OBLIGATIONS OF THE NATIONAL COMPETENT AUTHORITIES

Article 43

INFORMATION TO BE PROVIDED BY NATIONAL COMPETENT AUTHORITIES

National competent authorities shall make the following information available to the marketing authorisation holders, manufacturers, wholesalers and persons authorised or entitled to supply medicinal products to the public, upon their request:

(a) the medicinal products placed on the market on their territory which shall bear the safety features in accordance with Article 54(o) of Directive 2001/83/EC and this Regulation;
(b) the medicinal products subject to prescription or subject to reimbursement for which the scope of the unique identifier is extended for the purposes of reimbursement or pharmacovigilance, in accordance with Article 54a(5) of Directive 2001/83/EC;
(c) the medicinal products for which the scope of the anti-tampering device is extended for the purpose of patient safety, in accordance with Article 54a(5) of Directive 2001/83/EC.

Article 44

SUPERVISION OF THE REPOSITORIES SYSTEM

(1) National competent authorities shall supervise the functioning of any repository physically located in their territory, in order to verify, if necessary by means of inspections, that the repository and the legal entity responsible for the establishment and management of the repository comply with the requirements of this Regulation.
(2) A national competent authority may delegate any of its obligations under this Article to the competent authority of another Member State or to a third party, by means of a written agreement.
(3) Where a repository not physically located in the territory of a Member State is used for the purpose of verifying the authenticity of medicinal products placed on the market in that Member State, the competent authority of that Member State may observe an inspection of the repository or perform an independent inspection, subject to the agreement of the Member State in which the repository is physically located.

(4) A national competent authority shall communicate reports of supervision activities to the European Medicines Agency, which shall make them available to the other national competent authorities and the Commission.

(5) National competent authorities may contribute to the management of any repository used to identify medicinal products and verify the authenticity of or decommission the unique identifiers of medicinal products placed on the market in the territory of their Member State.

National competent authorities may participate to the management board of the legal entities managing those repositories to the extent of up to one third of the members of the board.

CHAPTER X: LISTS OF DEROGATIONS AND NOTIFICATIONS TO THE COMMISSION

Article 45

LISTS OF DEROGATIONS FROM BEARING OR NOT BEARING THE SAFETY FEATURES

(1) The list of medicinal products or product categories subject to prescription which shall not bear the safety features are set out in Annex I to this Regulation.

(2) The list of medicinal products or product categories not subject to prescription which shall bear the safety features are set out in Annex II to this Regulation.

Article 46

NOTIFICATIONS TO THE COMMISSION

(1) National competent authorities shall notify the Commission of non-prescription medicinal products which they judge to be at risk of falsification as soon as they become aware of such risk. For that purpose, they shall use the form set out in Annex III to this Regulation.

(2) National competent authorities may inform the Commission of medicinal products which they deem not to be at risk of falsification. For that purpose, they shall use the form set out in Annex IV to this Regulation.

(3) For the purposes of the notifications referred to in paragraphs 1 and 2, national competent authorities shall conduct an assessment of the risks of and arising from falsification of such products taking into account the criteria listed in Article 54a(2)(b) of Directive 2001/83/EC.

(4) When submitting to the Commission the notification referred to in paragraph 1, national competent authorities shall provide the Commission with evidence and documentation supporting the presence of incidents of falsification.

Article 47

EVALUATION OF THE NOTIFICATIONS

Where, following a notification as referred to in Article 46, the Commission or a Member State considers, on the basis of casualties or hospitalisations of citizens of the Union due to exposure to falsified medicinal products, that rapid action is required to protect public health, the Commission shall assess the notification without delay and at the latest within 45 days.

CHAPTER XI: TRANSITIONAL MEASURES AND ENTRY INTO FORCE

Article 48

TRANSITIONAL MEASURES

Medicinal products that have been released for sale or distribution without the safety features in a Member State before the date in which this Regulation becomes applicable in that Member State, and are not repackaged or relabelled thereafter, may be placed on the market, distributed and supplied to the public in that Member State until their expiry date.

Article 49

APPLICATION IN MEMBER STATES WITH EXISTING SYSTEMS FOR THE VERIFICATION OF THE AUTHENTICITY OF MEDICINAL PRODUCTS AND FOR THE IDENTIFICATION OF INDIVIDUAL PACKS

(1) Each of the Member States referred to in Article 2, paragraph 2, second subparagraph, point (b), second sentence, of Directive 2011/62/EU shall notify the Commission of the date from which Articles 1 to 48 of this Regulation apply in its territory in accordance with the third subparagraph of Article 50. The notification shall take place at the latest 6 months before that application.

(2) The Commission shall publish a notice of each of the dates notified to it in accordance with paragraph 1 in the *Official Journal of the European Union*.

Article 50

ENTRY INTO FORCE

This Regulation shall enter into force on the twentieth day following that of its publication in the *Official Journal of the European Union*.

It shall apply from 9 February 2019.

However, the Member States referred to in Article 2, paragraph 2, second subparagraph, point (b), second sentence, of Directive 2011/62/EU shall apply Articles 1 to 48 of this Regulation at the latest from 9 February 2025.

This Regulation shall be binding in its entirety and directly applicable in all Member States.

ANNEX I

List of medicinal products or product categories subject to prescription that shall not bear the safety features, referred to in Article 45(1)

Name of active substance or product category	Pharmaceutical form	Strength	Remarks
Homeopathic medicinal products	Any	Any	
Radionuclide generators	Any	Any	
Kits	Any	Any	
Radionuclide precursors	Any	Any	
Advanced therapy medicinal products which contain or consist of tissues or cells	Any	Any	
Medicinal gases	Medicinal gas	Any	
Solutions for parenteral nutrition having an anatomical therapeutical chemical ('ATC') code beginning with B05BA	Solution for infusion	Any	
Solutions affecting the electrolyte balance having an ATC code beginning with B05BB	Solution for infusion	Any	
Solutions producing osmotic diuresis having an ATC code beginning with B05BC	Solution for infusion	Any	
Intravenous solution additives having an ATC code beginning with B05X	Any	Any	
Solvents and diluting agents, including irrigating solutions, having an ATC code beginning with V07AB	Any	Any	

Name of active substance or product category	Pharmaceutical form	Strength	Remarks
Contrast media having an ATC code beginning with V08	Any	Any	
Tests for allergic diseases having an ATC code beginning with V04CL	Any	Any	
Allergen extracts having an ATC code beginning with V01AA	Any	Any	

ANNEX II

List of medicinal products or product categories not subject to prescription that shall bear the safety features, referred to in Article 45(2)

Name of active substance or product category	Pharmaceutical form	Strength	Remarks
omeprazole	gastro-resistant capsule, hard	20 mg	
omeprazole	gastro-resistant capsule, hard	40 mg	

ANNEX III

Notification to the European Commission of medicinal products not subject to prescription judged to be at risk of falsification, pursuant to article 54a(4) of Directive 2001/83/EC

Member State:			Name of competent authority:		
Entry No	**Active substance (Common Name)**	**Pharmaceutical form**	**Strength**	**Anatomical Therapeutical Chemical (ATC) Code**	**Supporting Evidence (please provide evidence of one or more incidents of falsification in the legal supply chain and specify the source of the information).**
1					
2					
3					
4					
5					
6					
7					
8					
9					
10					
11					
12					
13					
14					
15					

Note: The number of entries is not binding.

ANNEX IV

Notification to the European Commission of medicinal products judged not to be at risk of falsification, pursuant to article 54a(4) of Directive 2001/83/EC					
Member State:			**Name of competent authority:**		
Entry No	**Active substance (Common Name)**	**Pharmaceutical form**	**Strength**	**Anatomical Therapeutical Chemical (ATC) Code**	**Comments/ Complementary information**
1					
2					
3					
4					
5					
6					
7					
8					
9					
10					
11					
12					
13					
14					
15					

Note: The number of entries is not binding.

CHAPTER

3

UK Legislation on Wholesale Distribution

Contents

The Human Medicines Regulations 2012 [SI 2012/1916] 64
Citation and commencement 65
Medicinal products 65
General interpretation 65
Wholesale dealing in medicinal products 68
Exemptions from requirement for wholesale dealer's licence 69
Application for manufacturer's or wholesale dealer's licence 70
Factors relevant to determination of application for manufacturer's or wholesale dealer's licence 70
Grant or refusal of licence 71
Standard provisions of licences 72
Duration of licence 72
Conditions for wholesale dealer's licence 72
Obligations of licence holder 72
Requirement that wholesale dealers to deal only with specified persons 75
Requirement as to responsible persons 77
Schedule 4 Standard provisions of licences 78
Schedule 6 Manufacturer's and wholesale dealer's licences for exempt advanced therapy medicinal products 80

The Human Medicines Regulations 2012 [SI 2012/1916]

Editor's note These extracts from the Regulations and Standard Provisions of the Human Medicines Regulations 2012 [SI 2012/1916] are presented for the reader's convenience. Reproduction is with the permission of HMSO and the Queen's Printer for Scotland. For any definitive information reference must be made to the original Regulations. The numbering and content within this section corresponds with the regulations set out in the published Statutory Instrument [SI 2012/1916].

Citation and commencement

1 (1) These Regulations may be cited as the Human Medicines Regulations 2012.
(2) These Regulations come into force on 14th August 2012.

Medicinal products

2 (1) In these Regulations "medicinal product" means–
(a) any substance or combination of substances presented as having properties of preventing or treating disease in human beings; or
(b) any substance or combination of substances that may be used by or administered to human beings with a view to–
(i) restoring, correcting or modifying a physiological function by exerting a pharmacological, immunological or metabolic action, or
(ii) making a medical diagnosis.

General interpretation

8 (1) In these Regulations (unless the context otherwise requires):
"the 2001 Directive" means Directive 2001/83/EC of the European Parliament and of the Council on the Community Code relating to medicinal products for human use;
"Article 126a authorisation" means an authorisation granted by the licensing authority under Part 8 of these Regulations;
"brokering" means all activities in relation to the sale or purchase of medicinal products, except for wholesale distribution, that do not include physical handling and that consist of negotiating independently and on behalf of another legal or natural person;
"Directive 2002/98/EC" means Directive 2002/98/EC of the European Parliament and of the Council of 27 January 2003 setting standards of quality and safety for the collection, testing, processing, storage and distribution of human blood and blood components and amending Directive 2001/83/EC;
"Directive 2004/23/EC" means Directive 2004/23/EC of the European Parliament and of the Council of 31 March 2004 on setting standards of quality and safety for the donation, procurement, testing, processing, preservation, storage and distribution of human tissues and cells;

"electronic communication" means a communication transmitted (whether from one person to another, from one device to another or from a person to a device or vice versa):

(a) by means of an electronic communications network within the meaning of section 32(1) of the Communications Act 2003; or

(b) by other means but while in an electronic form;

"EU marketing authorisation" means a marketing authorisation granted or renewed by the European Commission under Regulation (EC) No 726/2004;

"European Economic Area" or "EEA" means the European Economic Area created by the EEA agreement;

"exempt advanced therapy medicinal product" has the meaning given in regulation 171;

"export" means export, or attempt to export, from the United Kingdom, whether by land, sea or air;

"falsified medicinal product" means any medicinal product with a false representation of:

(a) its identity, including its packaging and labelling, its name or its composition (other than any unintentional quality defect) as regards any of its ingredients including excipients and the strength of those ingredients;

(b) its source, including its manufacturer, its country of manufacturing, its country of origin or its marketing authorisation holder; or

(c) its history, including the records and documents relating to the distribution channels used;

"Fees Regulations" means the Medicines (Products for Human Use) (Fees) Regulations 2013[1];

"herbal medicinal product" means a medicinal product whose only active ingredients are herbal substances or herbal preparations (or both);

"herbal preparation" means a preparation obtained by subjecting herbal substances to processes such as extraction, distillation, expression, fractionation, purification, concentration or fermentation, and includes a comminuted or powdered herbal substance, a tincture, an extract, an essential oil, an expressed juice or a processed exudate;

"herbal substance" means a plant or part of a plant, algae, fungi or lichen, or an unprocessed exudate of a plant, defined by the plant part used and the botanical name of the plant, either fresh or dried, but otherwise unprocessed;

[1] S.I. 2013/532.

"homoeopathic medicinal product" means a medicinal product prepared from homoeopathic stocks in accordance with a homoeopathic manufacturing procedure described by:

(**a**) the European Pharmacopoeia; or

(**b**) in the absence of such a description in the European Pharmacopoeia, in any pharmacopoeia used officially in an EEA State;

"import" means import, or attempt to import, into the UK, whether by land, sea or air;

"inspector" means a person authorised in writing by an enforcement authority for the purposes of Part 16 (enforcement) (and references to "the enforcement authority", in relation to an inspector, are to the enforcement authority by whom the inspector is so authorised);

"the licensing authority" has the meaning given by regulation 6(2);

"manufacturer's licence" has the meaning given by regulation 17(1);

"marketing authorisation" means:

(**a**) a UK marketing authorisation; or

(**b**) an EU marketing authorisation;

"medicinal product subject to general sale" has the meaning given in regulation 5(1) (classification of medicinal products);

"Regulation (EC) No 726/2004" means Regulation (EC) No 726/2004 of the European Parliament and of the Council of 31 March 2004 laying down Community procedures for the authorisation and supervision of medicinal products for human and veterinary use and establishing a European Medicines Agency;

"Regulation (EC) No 1394/2007" means Regulation (EC) No 1394/2007 of the European Parliament and of the Council of 13 November 2007 on advanced therapy medicinal products and amending Directive 2001/83/EC and Regulation (EC) No 726/2004;

"Regulation (EC) No 1234/2008" means Commission Regulation (EC) No 1234/2008 of 24 November 2008 concerning the examination of variations to the terms of marketing authorisations for medicinal products for human use and veterinary medicinal products;

"the relevant EU provisions" means the provisions of legislation of the European Union relating to medicinal products for human use, except to the extent that any other enactment provides for any function in relation to any such provision to be exercised otherwise than by the licensing authority;

"relevant European State" means an EEA State or Switzerland;

"relevant medicinal product" has the meaning given by regulation 48;

"special medicinal product" means a product within the meaning of regulation 167 or any equivalent legislation in an EEA State other than the UK;

"third country" means a country or territory outside the EEA:

"traditional herbal medicinal product" means a herbal medicinal product to which regulation 125 applies;

"traditional herbal registration" means a traditional herbal registration granted by the licensing authority under these Regulations;

"UK marketing authorisation" means a marketing authorisation granted by the licensing authority under:

(a) Part 5 of these Regulations; or

(b) Chapter 4 of Title III to the 2001 Directive (mutual recognition and decentralised procedure);

"wholesale dealer's licence" has the meaning given by regulation 18(1).

(2) In these Regulations, references to distribution of a product by way of wholesale dealing are to be construed in accordance with regulation 18(4) and (5).

(3) In these Regulations, references to selling by retail, or to retail sale, are references to selling a product to a person who buys it otherwise than for a purpose specified in regulation 18(5).

(4) In these Regulations, references to supplying anything in circumstances corresponding to retail sale are references to supplying it, otherwise than by way of sale, to a person who receives it otherwise than for a purpose specified in regulation 18(5);

Wholesale dealing in medicinal products

18 (1) A person may not except in accordance with a licence (a "wholesale dealer's licence")–

(a) distribute a medicinal product by way of wholesale dealing; or

(b) possess a medicinal product for the purpose of such distribution.

(2) Paragraph (1)–

(a) does not apply–

(i) to anything done in relation to a medicinal product by the holder of a manufacturer's licence in respect of that product,

(ii) where the product concerned is an investigational medicinal product, or

(iii) if the product is a radiopharmaceutical in which the radionuclide is in the form of a sealed source; and

(b) is subject to regulation 19.

(3) Distribution of a medicinal product by way of wholesale dealing, or possession for the purpose of such distribution, is not to be taken to be in accordance with a wholesale dealer's licence unless the distribution

is carried on, or as the case may be the product held, at premises located in the UK and specified in the licence.

(4) In these Regulations a reference to distributing a product by way of wholesale dealing is a reference to–
- (a) selling or supplying it; or
- (b) procuring or holding it or exporting it for the purposes of sale or supply, to a person who receives it for a purpose within paragraph (5).

(5) Those purposes are–
- (a) selling or supplying the product; or
- (b) administering it or causing it to be administered to one or more human beings, in the course of a business carried on by that person.

(6) A wholesale dealer's licence does not authorise the distribution of a medicinal product by way of wholesale dealing, or possession for the purpose of such distribution, unless a marketing authorisation, Article 126a authorisation, certificate of registration or traditional herbal registration is in force in respect of the product but this–
- (a) . . .
- (b) is subject to the exceptions in regulation 43(6).

(7) In paragraph (6), "marketing authorisation" means–
- (a) a marketing authorisation issued by a competent authority of a member State in accordance with the 2001 Directive; or
- (b) an EU marketing authorisation.

Exemptions from requirement for wholesale dealer's licence

19 (1) Regulation 18 does not apply to the sale or offer for sale of a medicinal product by way of wholesale dealing, or possession for the purpose of such sale or offer, where paragraph (2) applies and the person selling or offering the product for sale is–
- (a) the holder of a marketing authorisation, Article 126a authorisation, certificate of registration or traditional herbal registration, (an "authorisation") which relates to the product, including a holder of an authorisation who manufactured or assembled the product; or
- (b) a person who is not the holder of an authorisation in relation to the product but manufactured or assembled the product to the order of a person who is the holder of an authorisation relating to the product.

(2) This paragraph applies if–
(a) until the sale, the medicinal product has been kept on the premises of the person who manufactured or assembled the product (in this regulation referred to as "authorised premises"); and
(b) those premises are premises authorised for use for manufacture or assembly by that person's manufacturer's licence.

(3) For the purposes of this regulation, a medicinal product is regarded as having been kept on authorised premises at a time when–
(a) it was being moved from one set of authorised premises to another, or from one part of authorised premises to another part; or
(b) it was being moved from authorised premises by way of delivery to a purchaser.

(4) Regulation 18 does not apply to a person who in connection with the importation of a medicinal product–
(a) provides facilities solely for transporting the product; or
(b) acting as an import agent, handles the product where the product is imported solely to the order of another person who intends to sell the product or offer it for sale by way of wholesale dealing or to distribute it in any other way.

Application for manufacturer's or wholesale dealer's licence

21 (1) An application for a grant of a licence under this Part must–
(a) be made to the licensing authority;
(b) be made in the way and form specified in Schedule 3; and
(c) contain or be accompanied by the information, documents, samples and other material specified in that Schedule.

(2) An application must indicate the descriptions of medicinal products in respect of which the licence is required, either by specifying the descriptions of medicinal products in question or by way of an appropriate general classification.

Factors relevant to determination of application for manufacturer's or wholesale dealer's licence

22 (1) In dealing with an application for a manufacturer's licence the licensing authority must in particular take into consideration–
(a) the operations proposed to be carried out under the licence;
(b) the premises in which those operations are to be carried out;
(c) the equipment which is or will be available on those premises for carrying out those operations;

(d) the qualifications of the persons under whose supervision the operations will be carried out; and
(e) the arrangements made or to be made for securing the safekeeping of, and the maintenance of adequate records in respect of, medicinal products manufactured or assembled in pursuance of the licence.

(2) In dealing with an application for a wholesale dealer's licence the licensing authority must in particular take into consideration–
(a) the premises on which medicinal products of the descriptions to which the application relates will be stored;
(b) the equipment which is or will be available for storing medicinal products on those premises;
(c) the equipment and facilities which are or will be available for distributing medicinal products from those premises; and
(d) the arrangements made or to be made for securing the safekeeping of, and the maintenance of adequate records in respect of, medicinal products stored on or distributed from those premises.

Grant or refusal of licence

23 (1) Subject to the following provisions of these Regulations, on an application to the licensing authority for a licence under this Part the licensing authority may–
(a) grant a licence containing such provisions as it considers appropriate; or
(b) refuse to grant a licence if having regard to the provisions of these Regulations and any European Union obligation it considers it necessary or appropriate to do so.

(2) The licensing authority must grant or refuse an application for a licence under this Part within the period of 90 days beginning immediately after the day on which it receives the application.

(3) Paragraph (2) applies to an application only if the requirements of Schedule 3 have been met.

(4) If a notice under regulation 30 requires the applicant to provide the licensing authority with information, the information period is not to be counted for the purposes of paragraph (2).

(5) In paragraph (4), the "information period" means the period–
(a) beginning with the day on which the notice is given, and
(b) ending with the day on which the licensing authority receives the information or the applicant shows to the licensing authority's satisfaction that the applicant is unable to provide it.

(6) The licensing authority must give the applicant a notice stating the reasons for its decision in any case where–

(a) the licensing authority refuses to grant an application for a licence; or

(b) the licensing authority grants a licence otherwise than in accordance with the application and the applicant requests a statement of its reasons.

Standard provisions of licences

24 (1) The standard provisions set out in Schedule 4 may be incorporated by the licensing authority in a licence under this Part granted on or after the date on which these Regulations come into force.

(2) The standard provisions may be incorporated in a licence with or without modifications and either generally or in relation to medicinal products of a particular class.

Duration of licence

25 A licence granted under this Part remains in force until–

(a) the licence is revoked by the licensing authority; or

(b) the licence is surrendered by the holder.

Conditions for wholesale dealer's licence

42 (1) Regulations 43 to 45 apply to the holder of a wholesale dealer's licence (referred to in those regulations as "the licence holder") and have effect as if they were provisions of the licence (but the provisions specified in paragraph (2) do not apply to the holder of a wholesale dealer's licence insofar as the licence relates to exempt advanced therapy medicinal products).

(2) Those provisions are regulations 43(2) and (8) and 44.

(3) The requirements in Part 2 of Schedule 6 apply to the holder of a wholesale dealer's licence insofar as the licence relates to exempt advanced therapy medicinal products, and have effect as if they were provisions of the licence.

Obligations of licence holder

43 (1) The licence holder must comply with the guidelines on good distribution practice published by the European Commission in accordance with Article 84 of the 2001 Directive.

(2) The licence holder must ensure, within the limits of the holder's responsibility, the continued supply of medicinal products to pharmacies, and other persons who may lawfully sell medicinal products by retail or supply them in circumstances corresponding to retail sale, so that the needs of patients in the United Kingdom are met.

(3) The licence holder must provide and maintain such staff, premises, equipment and facilities for the handling, storage and distribution of medicinal products under the licence as are necessary:
  (a) to maintain the quality of the products; and
  (b) to ensure their proper distribution.

(4) The licence holder must inform the licensing authority of any proposed structural alteration to, or discontinuance of use of, premises to which the licence relates or which have otherwise been approved by the licensing authority.

(5) Subject to paragraph (6), the licence holder must not sell or supply a medicinal product, or offer it for sale or supply, unless:
  (a) there is a marketing authorisation, Article 126a authorisation, certificate of registration or traditional herbal registration (an "authorisation") in force in relation to the product; and
  (b) the sale or supply, or offer for sale or supply, is in accordance with the authorisation.

(6) The restriction in paragraph (5) does not apply to:
  (a) the sale or supply, or offer for sale or supply, of a special medicinal product;
  (b) the export to an EEA State, or supply for the purposes of such export, of a medicinal product which may be placed on the market in that State without a marketing authorisation, Article 126a authorisation, certificate of registration or traditional herbal registration by virtue of legislation adopted by that State under Article 5(1) of the 2001 Directive;
  (c) the sale or supply, or offer for sale or supply, of an unauthorised medicinal product where the Secretary of State has temporarily authorised the distribution of the product under regulation 174; or
  (d) the wholesale distribution of medicinal products to a person in a third country.

(7) The licence holder must:
  (a) keep documents relating to the sale or supply of medicinal products under the licence which may facilitate the withdrawal or recall from sale of medicinal products in accordance with paragraph (b);
  (b) maintain an emergency plan to ensure effective implementation of the recall from the market of a medicinal product where recall is:
    (i) ordered by the licensing authority or by the competent authority of any EEA State, or

(ii) carried out in co-operation with the manufacturer of, or the holder of the marketing authorisation, Article 126a authorisation, certificate of registration or traditional herbal registration for, the product; and

(c) keep records in relation to the receipt, dispatch or brokering of medicinal products, of:

(i) the date of receipt,

(ii) the date of despatch,

(iii) the date of brokering,

(iv) the name of the medicinal product,

(v) the quantity of the product received, dispatched or brokered,

(vi) the name and address of the person from whom the products were received or to whom they are dispatched,

(vii) the batch number of medicinal products bearing safety features referred to in point (o) of Article 54[2] of the 2001 Directive.

(8) A licence holder ("L") who imports from another EEA State a medicinal product in relation to which L is not the holder of a marketing authorisation, Article 126a authorisation, certificate of registration or a traditional herbal registration shall:

(a) notify the intention to import that product to the holder of the authorisation and:

(i) in the case of a product which has been granted a marketing authorisation under Regulation (EC) No 726/2004, to the EMA; or

(ii) in any other case, the licensing authority; and

(b) pay a fee to the EMA in accordance with Article 76(4)[3] of the 2001 Directive or the licensing authority as the case may be, in accordance with the Fees Regulations, but this paragraph does not apply in relation to the wholesale distribution of medicinal products to a person in a third country.

(9) For the purposes of enabling the licensing authority to determine whether there are grounds for suspending, revoking or varying the licence, the licence holder must permit a person authorised in writing by the licensing authority, on production of identification, to carry out any inspection, or to take any samples or copies, which an inspector could carry out or take under Part 16 (enforcement).

---

[2] Point (o) of Article 54a was inserted by Directive 2011/62/EU of the European Parliament and of the Council (OJ No L 174, 1.7.2011, p74).

[3] Article 76(4) was inserted by Directive 2011/62/EU of the European Parliament and of the Council (OJ No L 174, 1.7.2011, p74).

(10) The holder ("L") must verify in accordance with paragraph (11) that any medicinal products received by L that are required by Article 54a[4] of the Directive to bear safety features are not falsified but this paragraph does not apply in relation to the distribution of medicinal products received from a third country by a person to a person in a third country.

(11) Verification under this paragraph is carried out by checking the safety features on the outer packaging, in accordance with the requirements laid down in the delegated acts adopted under Article 54a(2) of the 2001 Directive.

(12) The licence holder must maintain a quality system setting out responsibilities, processes and risk management measures in relation to their activities.

(13) The licence holder must immediately inform the licensing authority and, where applicable, the marketing authorisation holder, of medicinal products which the licence holder receives or is offered which the licence holder:

(a) knows or suspects; or

(b) has reasonable grounds for knowing or suspecting, to be falsified.

(14) Where the medicinal product is obtained through brokering, the licence holder must verify that the broker involved fulfils the requirements set out in regulation 45A(1)(b).

(15) In this regulation, "marketing authorisation" means:

(a) a marketing authorisation issued by a competent authority in accordance with the 2001 Directive; or

(b) an EU marketing authorisation.

Requirement that wholesale dealers to deal only with specified persons

44 (1) …

(2) The licence holder must not obtain supplies of medicinal products from anyone except:

(a) the holder of a manufacturer's licence or wholesale dealer's licence in relation to products of that description;

(b) the person who holds an authorisation granted by another EEA State authorising the manufacture of products of the description or their distribution by way of wholesale dealing; or

(c) where the medicinal product is directly received from a third country ("A") for export to a third country ("B"), the supplier of

[4] Article 54a was inserted by Directive 2011/62/EU of the European Parliament and of the Council (OJ No L 174, 1.7.2011, p74).

the medicinal product in country A is a person who is authorised or entitled to supply such medicinal products in accordance with the legal and administrative provisions in country A.

(3) Where a medicinal product is obtained in accordance with paragraph (2)(a) or (b), the licence holder must verify that:
  (a) the wholesale dealer who supplies the product complies with the principles and guidelines of good distribution practices; or
  (b) the manufacturer or importer who supplies the product holds a manufacturing authorisation.

(4) ...

(5) The licence holder may distribute medicinal products by way of wholesale dealing only to:
  (a) the holder of a wholesale dealer's licence relating to those products;
  (b) the holder of an authorisation granted by the competent authority of another EEA State authorising the supply of those products by way of wholesale dealing;
  (c) a person who may lawfully sell those products by retail or may lawfully supply them in circumstances corresponding to retail sale;
  (d) a person who may lawfully administer those products; or
  (e) in relation to supply to persons in third countries, a person who is authorised or entitled to receive medicinal products for wholesale distribution or supply to the public in accordance with the applicable legal and administrative provisions of the third country concerned.

(6) Where a medicinal product is supplied to a person who is authorised or entitled to supply medicinal products to the public in accordance with paragraph (5)(c) or (e), the licence holder must enclose with the product a document stating the:
  (a) date on which the supply took place;
  (b) name and pharmaceutical form of the product supplied;
  (c) quantity of product supplied;
  (d) name and address of the licence holder; and
  (e) batch number of the medicinal products bearing the safety features referred to in point (o) of Article 54 of the 2001 Directive.

(7) The licence holder must:
  (a) keep a record of information supplied in accordance with paragraph (6) for at least five years beginning immediately after the date on which the information is supplied; and
  (b) ensure that the record is available to the licensing authority for inspection.

Requirement as to responsible persons

45 (1) The licence holder must ensure that there is available at all times at least one person (referred to in this regulation as the "responsible person") who in the opinion of the licensing authority:

(a) has knowledge of the activities to be carried out and of the procedures to be performed under the licence which is adequate to carry out the functions mentioned in paragraph (2); and

(b) has adequate experience relating to those activities and procedures.

(2) Those functions are:

(a) ensuring that the conditions under which the licence was granted have been, and are being, complied with; and

(b) ensuring that the quality of medicinal products handled by the licence holder is being maintained in accordance with the requirements of the marketing authorisations, Article 126a authorisations, certificates of registration or traditional herbal registrations applicable to those products.

(3) The licence holder must notify the licensing authority of:

(a) any change to the responsible person; and

(b) the name, address, qualifications and experience of the responsible person.

(4) The licence holder must not permit any person to act as a responsible person other than the person named in the licence or another person notified to the licensing authority under paragraph (3).

(5) Paragraph (6) applies if, after giving the licence holder and a person acting as a responsible person the opportunity to make representations (orally or in writing), the licensing authority thinks that the person:

(a) does not satisfy the requirements of paragraph (1) in relation to qualifications or experience; or

(b) is failing to carry out the functions referred to in paragraph (2) adequately or at all.

(6) Where this paragraph applies, the licensing authority must notify the licence holder in writing that the person is not permitted to act as a responsible person.

Schedule 4 Standard provisions of licences

PART 4 WHOLESALE DEALER'S LICENCE

All wholesale dealer's licences

28 The provisions of this Part are standard provisions of a wholesale dealer's licence.

29 The licence holder must not use any premises for the handling, storage or distribution of medicinal products other than those specified in the licence or notified to the licensing authority from time to time and approved by the licensing authority.

30 The licence holder must provide such information as may be requested by the licensing authority concerning the type and quantity of medicinal products which the licence holder handles, stores or distributes.

31 The licence holder must take all reasonable precautions and exercise all due diligence to ensure that any information provided by the licence holder to the licensing authority which is relevant to an evaluation of the safety, quality or efficacy of a medicinal product which the licence holder handles, stores or distributes is not false or misleading.

Wholesale dealer's licence relating to special medicinal products

32 The provisions of paragraphs 33 to 42 are incorporated as additional standard provisions of a wholesale dealer's licence relating to special medicinal products.

33 Where and in so far as the licence relates to special medicinal products, the licence holder may only import such products from another EEA State:

(a) in response to an order which satisfies the requirements of regulation 167, and
(b) where the conditions set out in paragraphs 34 to 41 are complied with.

34 No later than 28 days prior to each importation of a special medicinal product, the licence holder must give written notice to the licensing authority stating the intention to import the product and stating the following particulars:

(a) the brand name, common name or scientific name of the medicinal product and (if different) any name under which the medicinal product is to be sold or supplied in the United Kingdom;
(b) any trademark or the name of the manufacturer of the medicinal product;
(c) in respect of each active constituent of the medicinal product, any international non-proprietary name or the British approved name or

the monograph name, or where that constituent does not have any of those, the accepted scientific name or any other name descriptive of the true nature of the constituent;

(d) the quantity of medicinal product to be imported, which must not exceed the quantity specified in paragraph 38; and

(e) the name and address of the manufacturer or assembler of the medicinal product in the form in which it is to be imported and, if the person who will supply the medicinal product for importation is not the manufacturer or assembler, the name and address of the supplier.

35 The licence holder may not import the special medicinal product if, before the end of 28 days beginning immediately after the date on which the licensing authority sends or gives the licence holder an acknowledgement in writing by the licensing authority that it has received the notice referred to in paragraph 34, the licensing authority has notified the licence holder in writing that the product should not be imported.

36 The licence holder may import the special medicinal product referred to in the notice where the licence holder has been notified in writing by the licensing authority, before the end of the 28-day period referred to in paragraph 35, that the product may be imported.

37 Where the licence holder sells or supplies special medicinal products, the licence holder must, in addition to any other records which are required by the provisions of the licence, make and maintain written records relating to:

(a) the batch number of the batch of the product from which the sale or supply was made; and

(b) details of any adverse reaction to the product sold or supplied of which the licence holder becomes aware.

38 The licence holder must not, on any one occasion, import more than such amount as is sufficient for 25 single administrations, or for 25 courses of treatment where the amount imported is sufficient for a maximum of three months' treatment, and must not, on any one occasion, import more than the quantity notified to the licensing authority under paragraph 34(d).

39 The licence holder must inform the licensing authority immediately of any matter coming to the licence holder's attention which might reasonably cause the licensing authority to believe that a special medicinal product imported in accordance with this paragraph can no longer be regarded as a product which can safely be administered to human beings or as a product which is of satisfactory quality for such administration.

40 The licence holder must not publish any advertisement, catalogue, or circular relating to a special medicinal product or make any representations in respect of that product.

41 The licence holder must cease importing or supplying a special medicinal product if the licence holder receives a notice in writing from the licensing authority directing that, from a date specified in the notice, a particular product or class of products may no longer be imported or supplied.

42 In this Part:

"British approved name" means the name which appears in the current edition of the list prepared by the British Pharmacopoeia Commission under regulation 318 (British Pharmacopoeia- lists of names);
"international non-proprietary name" means a name which has been selected by the World Health Organisation as a recommended international non-proprietary name and in respect of which the Director-General of the World Health Organisation has given notice to that effect in the World Health Organisation Chronicle; and
"monograph name" means the name or approved synonym which appears at the head of a monograph in the current edition of the British Pharmacopoeia, the European Pharmacopoeia or a foreign or international compendium of standards, and "current" in this definition means current at the time the notice is sent to the licensing authority.

Wholesale dealer's licence relating to exempt advanced therapy medicinal products

43 The provisions of paragraph 44 are incorporated as additional standard provisions of a wholesale dealer's licence relating to exempt advanced therapy medicinal products.

44 The licence holder shall keep the data referred to in paragraph 16 of Schedule 6 for such period, being a period of longer than 30 years, as may be specified by the licensing authority.

Schedule 6 Manufacturer's and wholesale dealer's licences for exempt advanced therapy medicinal products

PART 2 WHOLESALE DEALER'S LICENCES

13 The requirements in paragraphs 14 to 20 apply to a wholesale dealer's licence insofar as it relates to exempt advanced therapy medicinal products.

14 The licence holder must obtain supplies of exempt advanced therapy medicinal products only from:

(**a**) the holder of a manufacturer's licence in respect of those products; or
(**b**) the holder of a wholesale dealer's licence in respect of those products.

15 The licence holder must distribute an exempt advanced therapy medicinal product by way of wholesale dealing only to:

(**a**) the holder of a wholesale dealer's licence in respect of those products; or
(**b**) a person who:
 (**i**) may lawfully administer those products, and
 (**ii**) solicited the product for an individual patient.

16 The licence holder must establish and maintain a system ensuring that the exempt advanced therapy medicinal product and its starting and raw materials, including all substances coming into contact with the cells or tissues it may contain, can be traced through the sourcing, manufacturing, packaging, storage, transport and delivery to the establishment where the product is used.

17 The licence holder must inform the licensing authority of any adverse reaction to any exempt advanced therapy medicinal product supplied by the holder of the wholesale dealer's licence of which the holder is aware.

18 The licence holder must, subject to paragraph 44 of Schedule 4, keep the data referred to in paragraph 16 for a minimum of 30 years after the expiry date of the exempt advanced therapy medicinal product.

19 The licence holder must secure that the data referred to in paragraph 16 will, in the event that:

(**a**) the licence is suspended, revoked or withdrawn; or
(**b**) the licence holder becomes bankrupt or insolvent,

be held available to the licensing authority by the holder of a wholesale dealer's licence for the period described in paragraph 18 or such longer period as may be required pursuant to paragraph 44 of Schedule 4.

20 The licence holder must not import or export any exempt advanced therapy medicinal product.

CHAPTER

4

EU Guidelines on Good Distribution Practice of Medicinal Products for Human Use (2013/C 343/01)

Contents

Guidelines on Good Distribution Practice of Medicinal Products for Human Use (2013/C 343/01) 83

Introduction 83
Chapter 1 – Quality Management 84
1.1 Principle 84
1.2 Quality system 84
1.3 Management of outsourced activities 85
1.4 Management review and monitoring 85
1.5 Quality risk management 86
Chapter 2 – Personnel 86
2.1 Principle 86
2.2 Responsible person 86
2.3 Other personnel 87
2.4 Training 88
2.5 Hygiene 88
Chapter 3 – Premises and Equipment 88
3.1 Principle 88
3.2 Premises 89
3.3 Equipment 90
Chapter 4 – Documentation 92
4.1 Principle 92
4.2 General 92
Chapter 5 – Operations 93
5.1 Principle 93
5.2 Qualification of suppliers 94
5.3 Qualification of customers 95
5.4 Receipt of medicinal products 95
5.5 Storage 95
5.6 Destruction of obsolete goods 96
5.7 Picking 96
5.8 Supply 96
5.9 Export to third countries 97
Chapter 6 – Complaints, Returns, Suspected Falsified Medicinal Products and Medicinal Product Recalls 97
6.1 Principle 97
6.2 Complaints 97
6.3 Returned medicinal products 98
6.4 Falsified medicinal products 99
6.5 Medicinal product recalls 99
Chapter 7 – Outsourced Activities 100
7.1 Principle 100
7.2 Contract giver 100
7.3 Contract acceptor 100
Chapter 8 – Self-inspections 101
8.1 Principle 101
8.2 Self-inspections 101
Chapter 9 – Transportation 101
9.1 Principle 101
9.2 Transportation 102
9.3 Containers, packaging and labelling 103
9.4 Products requiring special conditions 103
Chapter 10 – Specific Provisions for Brokers 104
10.1 Principle 104
10.2 Quality system 104
10.3 Personnel 105
10.4 Documentation 105
Chapter 11 – Final Provisions 105
Annex 106

European Commission Q&A on GDP Guidelines 108

GUIDELINES ON GOOD DISTRIBUTION PRACTICE OF MEDICINAL PRODUCTS FOR HUMAN USE (2013/C 343/01)

Introduction

These Guidelines are based on Article 84 and Article 85b(3) of Directive 2001/83/EC[1].

The Commission has published EU Guidelines on Good Distribution Practice (GDP) in 1994[2]. Revised guidelines were published in March 2013[3] in order to take into account recent advances in practices for appropriate storage and distribution of medicinal products in the European Union, as well as new requirements introduced by Directive 2011/62/EU[4].

This version corrects factual mistakes identified in subchapters 5.5 and 6.3 of the revised guidelines. It also gives more explanations on the rationale for the revision as well as a date of coming into operation.

It replaces the guidelines on GDP published in March 2013.

The wholesale distribution of medicinal products is an important activity in integrated supply chain management. Today's distribution network for medicinal products is increasingly complex and involves many players. These Guidelines lay down appropriate tools to assist wholesale distributors in conducting their activities and to prevent falsified medicines from entering the legal supply chain. Compliance with these Guidelines will ensure control of the distribution chain and consequently maintain the quality and the integrity of medicinal products.

According to Article 1(17) of Directive 2001/83/EC, wholesale distribution of medicinal products is 'all activities consisting of procuring, holding, supplying or exporting medicinal products, apart from supplying medicinal products to the public. Such activities are carried out with manufacturers or their depositories, importers, other wholesale distributors or with pharmacists and persons authorized or entitled to supply medicinal products to the public in the Member State concerned'.

---

1 Directive 2001/83/EC of the European Parliament and of the Council of 6 November 2001 on the Community code relating to medicinal products for human use, OJ L 311, 28.11.2001, p. 67.

2 Guidelines on Good Distribution Practice of medicinal products for human use, OJ C 63, 1.3.1994, p. 4.

3 Guidelines of 7 March 2013 on Good Distribution Practice of medicinal products for human use, OJ C 68, 8.3.2013, p. 1.

4 Directive 2011/62/EU of the European Parliament and of the Council amending Directive 2001/83/EC as regards the prevention of the entry into the legal supply chain of falsified medicinal products, OJ L 174, 1.7.2011, p. 74.

Any person acting as a wholesale distributor has to hold a wholesale distribution authorisation. Article 80(g) of Directive 2001/83/EC provides that distributors must comply with the principle of and guidelines for GDP.

Possession of a manufacturing authorisation includes authorisation to distribute the medicinal products covered by the authorisation. Manufacturers performing any distribution activities with their own products must therefore comply with GDP.

The definition of wholesale distribution does not depend on whether that distributor is established or operating in specific customs areas, such as in free zones or in free warehouses. All obligations related to wholesale distribution activities (such as exporting, holding or supplying) also apply to these distributors. Relevant sections of these Guidelines should also be adhered to by other actors involved in the distribution of medicinal products.

Other actors such as brokers may also play a role in the distribution channel for medicinal products. According to Article 85(b) of Directive 2001/83/EC, persons brokering medicinal products must be subject to certain provisions applicable to wholesale distributors, as well as specific provisions on brokering.

Chapter 1 – Quality Management

1.1 Principle

Wholesale distributors must maintain a quality system setting out responsibilities, processes and risk management principles in relation to their activities[1]. All distribution activities should be clearly defined and systematically reviewed. All critical steps of distribution processes and significant changes should be justified and where relevant validated. The quality system is the responsibility of the organisation's management and requires their leadership and active participation and should be supported by staff commitment.

1.2 Quality system

The system for managing quality should encompass the organisational structure, procedures, processes and resources, as well as activities necessary to ensure confidence that the product delivered maintains its quality and integrity and remains within the legal supply chain during storage and/or transportation.

The quality system should be fully documented and its effectiveness monitored. All quality system-related activities should be defined and

[1] Article 80(h) of Directive 2001/83/EC.

documented. A quality manual or equivalent documentation approach should be established.

A responsible person should be appointed by the management, who should have clearly specified authority and responsibility for ensuring that a quality system is implemented and maintained.

The management of the distributor should ensure that all parts of the quality system are adequately resourced with competent personnel, and suitable and sufficient premises, equipment and facilities.

The size, structure and complexity of distributor's activities should be taken into consideration when developing or modifying the quality system.

A change control system should be in place. This system should incorporate quality risk management principles, and be proportionate and effective.

The quality system should ensure that:

(i) medicinal products are procured, held, supplied or exported in a way that is compliant with the requirements of GDP;
(ii) management responsibilities are clearly specified;
(iii) products are delivered to the right recipients within a satisfactory time period;
(iv) records are made contemporaneously;
(v) deviations from established procedures are documented and investigated;
(vi) appropriate corrective and preventive actions (commonly known as 'CAPA') are taken to correct deviations and prevent them in line with the principles of quality risk management.

1.3 Management of outsourced activities

The quality system should extend to the control and review of any outsourced activities related to the procurement, holding, supply or export of medicinal products. These processes should incorporate quality risk management and include:

(i) assessing the suitability and competence of the contract acceptor to carry out the activity and checking authorisation status, if required;
(ii) defining the responsibilities and communication processes for the quality-related activities of the parties involved;
(iii) monitoring and review of the performance of the contract acceptor, and the identification and implementation of any required improvements on a regular basis.

1.4 Management review and monitoring

The management should have a formal process for reviewing the quality system on a periodic basis. The review should include:

(i) measurement of the achievement of quality system objectives;
(ii) assessment of performance indicators that can be used to monitor the effectiveness of processes within the quality system, such as complaints, deviations, CAPA, changes to processes; feedback on outsourced activities; self-assessment processes including risk assessments and audits; and external assessments such as inspections, findings and customer audits;
(iii) emerging regulations, guidance and quality issues that can impact the quality management system;
(iv) innovations that might enhance the quality system;
(v) changes in business environment and objectives.

The outcome of each management review of the quality system should be documented in a timely manner and effectively communicated internally.

1.5 Quality risk management

Quality risk management is a systematic process for the assessment, control, communication and review of risks to the quality of medicinal products. It can be applied both proactively and retrospectively.

Quality risk management should ensure that the evaluation of the risk to quality is based on scientific knowledge, experience with the process and ultimately links to the protection of the patient. The level of effort, formality and documentation of the process should be commensurate with the level of risk. Examples of the processes and applications of quality risk management can be found in guideline Q9 of the International Conference on Harmonisation ('ICH').

Chapter 2 – Personnel

2.1 Principle

The correct distribution of medicinal products relies upon people. For this reason, there must be sufficient competent personnel to carry out all the tasks for which the wholesale distributor is responsible. Individual responsibilities should be clearly understood by the staff and be recorded.

2.2 Responsible person

The wholesale distributor must designate a person as responsible person. The responsible person should meet the qualifications and all conditions provided for by the legislation of the Member State concerned[1]. A degree in

[1] Article 79(b) of Directive 2001/83/EC.

pharmacy is desirable. The responsible person should have appropriate competence and experience as well as knowledge of and training in GDP.

The responsible person should fulfil their responsibilities personally and should be continuously contactable. The responsible person may delegate duties but not responsibilities.

The written job description of the responsible person should define their authority to take decisions with regard to their responsibilities. The wholesale distributor should give the responsible person the defined authority, resources and responsibility needed to fulfil their duties.

The responsible person should carry out their duties in such a way as to ensure that the wholesale distributor can demonstrate GDP compliance and that public service obligations are met.

The responsibilities of the responsible person include:

(i) ensuring that a quality management system is implemented and maintained;
(ii) focusing on the management of authorised activities and the accuracy and quality of records;
(iii) ensuring that initial and continuous training programmes are implemented and maintained;
(iv) coordinating and promptly performing any recall operations for medicinal products;
(v) ensuring that relevant customer complaints are dealt with effectively;
(vi) ensuring that suppliers and customers are approved;
(vii) approving any subcontracted activities which may impact on GDP;
(viii) ensuring that self-inspections are performed at appropriate regular intervals following a prearranged programme and necessary corrective measures are put in place;
(ix) keeping appropriate records of any delegated duties;
(x) deciding on the final disposition of returned, rejected, recalled or falsified products;
(xi) approving any returns to saleable stock;
(xii) ensuring that any additional requirements imposed on certain products by national law are adhered to[2].

2.3 Other personnel

There should be an adequate number of competent personnel involved in all stages of the wholesale distribution activities of medicinal products. The number of personnel required will depend on the volume and scope of activities.

[2] Article 83 of Directive 2001/83/EC.

The organisational structure of the wholesale distributor should be set out in an organisation chart. The role, responsibilities, and interrelationships of all personnel should be clearly indicated.

The role and responsibilities of employees working in key positions should be set out in written job descriptions, along with any arrangements for deputising.

2.4 Training

All personnel involved in wholesale distribution activities should be trained on the requirements of GDP. They should have the appropriate competence and experience prior to commencing their tasks.

Personnel should receive initial and continuing training relevant to their role, based on written procedures and in accordance with a written training programme. The responsible person should also maintain their competence in GDP through regular training.

In addition, training should include aspects of product identification and avoidance of falsified medicines entering the supply chain.

Personnel dealing with any products which require more stringent handling conditions should receive specific training. Examples of such products include hazardous products, radioactive materials, products presenting special risks of abuse (including narcotic and psychotropic substances), and temperature-sensitive products.

A record of all training should be kept, and the effectiveness of training should be periodically assessed and documented.

2.5 Hygiene

Appropriate procedures relating to personnel hygiene, relevant to the activities being carried out, should be established and observed. Such procedures should cover health, hygiene and clothing.

Chapter 3 – Premises and Equipment

3.1 Principle

Wholesale distributors must have suitable and adequate premises, installations and equipment[1], so as to ensure proper storage and distribution of medicinal products. In particular, the premises should be clean, dry and maintained within acceptable temperature limits.

[1] Article 79(a) of Directive 2001/83/EC.

3.2 Premises

The premises should be designed or adapted to ensure that the required storage conditions are maintained. They should be suitably secure, structurally sound and of sufficient capacity to allow safe storage and handling of the medicinal products. Storage areas should be provided with adequate lighting to enable all operations to be carried out accurately and safely.

Where premises are not directly operated by the wholesale distributor, a contract should be in place. The contracted premises should be covered by a separate wholesale distribution authorisation.

Medicinal products should be stored in segregated areas which are clearly marked and have access restricted to authorised personnel. Any system replacing physical segregation, such as electronic segregation based on a computerised system, should provide equivalent security and should be validated.

Products pending a decision as to their disposition or products that have been removed from saleable stock should be segregated either physically or through an equivalent electronic system. This includes, for example, any product suspected of falsification and returned products. Medicinal products received from a third country but not intended for the Union market should also be physically segregated. Any falsified medicinal products, expired products, recalled products and rejected products found in the supply chain should be immediately physically segregated and stored in a dedicated area away from all other medicinal products. The appropriate degree of security should be applied in these areas to ensure that such items remain separate from saleable stock. These areas should be clearly identified.

Special attention should be paid to the storage of products with specific handling instructions as specified in national law. Special storage conditions (and special authorisations) may be required for such products (e.g. narcotics and psychotropic substances).

Radioactive materials and other hazardous products, as well as products presenting special safety risks of fire or explosion (e.g. medicinal gases, combustibles, flammable liquids and solids), should be stored in one or more dedicated areas subject to local legislation and appropriate safety and security measures.

Receiving and dispatch bays should protect products from prevailing weather conditions. There should be adequate separation between the receipt and dispatch and storage areas. Procedures should be in place to maintain control of inbound/outbound goods. Reception areas where deliveries are examined following receipt should be designated and suitably equipped.

Unauthorised access to all areas of the authorised premises should be prevented. Prevention measures would usually include a monitored intruder alarm system and appropriate access control. Visitors should be accompanied.

Premises and storage facilities should be clean and free from litter and dust. Cleaning programmes, instructions and records should be in place. Appropriate cleaning equipment and cleaning agents should be chosen and used so as not to present a source of contamination.

Premises should be designed and equipped so as to afford protection against the entry of insects, rodents or other animals. A preventive pest control programme should be in place.

Rest, wash and refreshment rooms for employees should be adequately separated from the storage areas. The presence of food, drink, smoking material or medicinal products for personal use should be prohibited in the storage areas.

3.2.1. TEMPERATURE AND ENVIRONMENT CONTROL

Suitable equipment and procedures should be in place to check the environment where medicinal products are stored. Environmental factors to be considered include temperature, light, humidity and cleanliness of the premises.

An initial temperature mapping exercise should be carried out on the storage area before use, under representative conditions. Temperature monitoring equipment should be located according to the results of the mapping exercise, ensuring that monitoring devices are positioned in the areas that experience the extremes of fluctuations. The mapping exercise should be repeated according to the results of a risk assessment exercise or whenever significant modifications are made to the facility or the temperature controlling equipment. For small premises of a few square meters which are at room temperature, an assessment of potential risks (e.g. heaters) should be conducted and temperature monitors placed accordingly.

3.3 Equipment

All equipment impacting on storage and distribution of medicinal products should be designed, located and maintained to a standard which suits its intended purpose. Planned maintenance should be in place for key equipment vital to the functionality of the operation.

Equipment used to control or to monitor the environment where the medicinal products are stored should be calibrated at defined intervals based on a risk and reliability assessment.

Calibration of equipment should be traceable to a national or international measurement standard. Appropriate alarm systems should be in place to provide alerts when there are excursions from pre-defined storage conditions. Alarm levels should be appropriately set and alarms should be regularly tested to ensure adequate functionality.

Equipment repair, maintenance and calibration operations should be carried out in such a way that the integrity of the medicinal products is not compromised.

Adequate records of repair, maintenance and calibration activities for key equipment should be made and the results should be retained. Key equipment would include for example cold stores, monitored intruder alarm and access control systems, refrigerators, thermo hygrometers, or other temperature and humidity recording devices, air handling units and any equipment used in conjunction with the onward supply chain.

3.3.1. COMPUTERISED SYSTEMS

Before a computerised system is brought into use, it should be demonstrated, through appropriate validation or verification studies, that the system is capable of achieving the desired results accurately, consistently and reproducibly.

A written, detailed description of the system should be available (including diagrams where appropriate). This should be kept up to date. The document should describe principles, objectives, security measures, system scope and main features, how the computerised system is used and the way it interacts with other systems.

Data should only be entered into the computerised system or amended by persons authorised to do so.

Data should be secured by physical or electronic means and protected against accidental or unauthorised modifications. Stored data should be checked periodically for accessibility. Data should be protected by backing up at regular intervals. Back up data should be retained for the period stated in national legislation but at least five years at a separate and secure location.

Procedures to be followed if the system fails or breaks down should be defined. This should include systems for the restoration of data.

3.3.2. QUALIFICATION AND VALIDATION

Wholesale distributors should identify what key equipment qualification and/or key process validation is necessary to ensure correct installation and operation. The scope and extent of such qualification and/or validation activities (such as storage, pick and pack processes) should be determined using a documented risk assessment approach.

Equipment and processes should be respectively qualified and/or validated before commencing use and after any significant changes e.g. repair or maintenance.

Validation and qualification reports should be prepared summarising the results obtained and commenting on any observed deviations. Deviations from established procedures should be documented and further actions decided to correct deviations and avoid their reoccurrence (corrective and preventive actions). The principles of CAPA should be applied where necessary. Evidence of satisfactory validation and acceptance of a process or piece of equipment should be produced and approved by appropriate personnel.

Chapter 4 – Documentation

4.1 Principle

Good documentation constitutes an essential part of the quality system. Written documentation should prevent errors from spoken communication and permits the tracking of relevant operations during the distribution of medicinal products.

4.2 General

Documentation comprises all written procedures, instructions, contracts, records and data, in paper or in electronic form. Documentation should be readily available/retrievable.

With regard to the processing of personal data of employees, complainants or any other natural person, Directive 95/46/EC[1] on the protection of individuals applies to the processing of personal data and to the free movement of such data.

Documentation should be sufficiently comprehensive with respect to the scope of the wholesale distributor's activities and in a language understood by personnel. It should be written in clear, unambiguous language and be free from errors.

Procedure should be approved signed and dated by the responsible person. Documentation should be approved, signed and dated by appropriate authorised persons, as required. It should not be handwritten; although, where it is necessary, sufficient space should be provided for such entries.

[1] OJ L 281, 23.11.1995, p. 31.

Any alteration made in the documentation should be signed and dated; the alteration should permit the reading of the original information. Where appropriate, the reason for the alteration should be recorded.

Documents should be retained for the period stated in national legislation but at least five years. Personal data should be deleted or anonymised as soon as their storage is no longer than necessary for the purpose of distribution activities.

Each employee should have ready access to all necessary documentation for the tasks executed.

Attention should be paid to using valid and approved procedures. Documents should have unambiguous content; title, nature and purpose should be clearly stated. Documents should be reviewed regularly and kept up to date. Version control should be applied to procedures. After revision of a document a system should exist to prevent inadvertent use of the superseded version. Superseded or obsolete procedures should be removed from workstations and archived.

Records must be kept either in the form of purchase/sales invoices, delivery slips, or on computer or any other form, for any transaction in medicinal products received, supplied or brokered.

Records must include at least the following information: date; name of the medicinal product; quantity received, supplied or brokered; name and address of the supplier, customer, broker or consignee, as appropriate; and batch number at least for medicinal product bearing the safety features[2].

Records should be made at the time each operation is undertaken.

Chapter 5 – Operations

5.1 Principle

All actions taken by wholesale distributors should ensure that the identity of the medicinal product is not lost and that the wholesale distribution of medicinal products is performed according to the information on the outer packaging. The wholesale distributor should use all means available to minimise the risk of falsified medicinal products entering the legal supply chain.

All medicinal products distributed in the EU by a wholesale distributor must be covered by a marketing authorisation granted by the EU or by a Member State[3].

Any distributor, other than the marketing authorisation holder, who imports a medicinal product from another Member State must notify the marketing authorisation holder and the competent authority in the

[2] Articles 80(e) and 82 of Directive 2001/83/EC.

[3] Articles 76(1) and (2) of Directive 2001/83/EC.

Member State to which the medicinal product will be imported of their intention to import that product[4]. All key operations described below should be fully described in the quality system in appropriate documentation.

5.2 Qualification of suppliers

Wholesale distributors must obtain their supplies of medicinal products only from persons who are themselves in possession of a wholesale distribution authorisation, or who are in possession of a manufacturing authorisation which covers the product in question[5].

Wholesale distributors receiving medicinal products from third countries for the purpose of importation, i.e. for the purpose of placing these products on the EU market, must hold a manufacturing authorisation[6].

Where medicinal products are obtained from another wholesale distributor the receiving wholesale distributor must verify that the supplier complies with the principles and guidelines of good distribution practices and that they hold an authorisation for example by using the Union database. If the medicinal product is obtained through brokering, the wholesale distributor must verify that the broker is registered and complies with the requirements in Chapter 10[1].

Appropriate qualification and approval of suppliers should be performed prior to any procurement of medicinal products. This should be controlled by a procedure and the results documented and periodically rechecked.

When entering into a new contract with new suppliers the wholesale distributor should carry out 'due diligence' checks in order to assess the suitability, competence and reliability of the other party. Attention should be paid to:

(i) the reputation or reliability of the supplier;
(ii) offers of medicinal products more likely to be falsified;
(iii) large offers of medicinal products which are generally only available in limited quantities; and
(iv) out-of-range prices.

[4] Article 76(3) of Directive 2001/83/EC.
[5] Article 80(b) of Directive 2001/83/EC.
[6] Article 40, third paragraph of Directive 2001/83/EC.
[1] Article 80, fourth paragraph of Directive 2001/83/EC.

5.3 Qualification of customers

Wholesale distributors must ensure they supply medicinal products only to persons who are themselves in possession of a wholesale distribution authorisation or who are authorised or entitled to supply medicinal products to the public.

Checks and periodic rechecks may include: requesting copies of customer's authorisations according to national law, verifying status on an authority website, requesting evidence of qualifications or entitlement according to national legislation.

Wholesale distributors should monitor their transactions and investigate any irregularity in the sales patterns of narcotics, psychotropic substances or other dangerous substances. Unusual sales patterns that may constitute diversion or misuse of medicinal product should be investigated and reported to competent authorities where necessary. Steps should be taken to ensure fulfilment of any public service obligation imposed upon them.

5.4 Receipt of medicinal products

The purpose of the receiving function is to ensure that the arriving consignment is correct, that the medicinal products originate from approved suppliers and that they have not been visibly damaged during transport.

Medicinal products requiring special storage or security measures should be prioritised and once appropriate checks have been conducted they should be immediately transferred to appropriate storage facilities.

Batches of medicinal products intended for the EU and EEA countries should not be transferred to saleable stock before assurance has been obtained in accordance with written procedures, that they are authorised for sale. For batches coming from another Member State, prior to their transfer to saleable stock, the control report referred to in Article 51(1) of Directive 2001/83/EC or another proof of release to the market in question based on an equivalent system should be carefully checked by appropriately trained personnel.

5.5 Storage

Medicinal products and, if necessary, healthcare products should be stored separately from other products likely to alter them and should be protected from the harmful effects of light, temperature, moisture and other external factors. Particular attention should be paid to products requiring specific storage conditions.

Incoming containers of medicinal products should be cleaned, if necessary, before storage.

Warehousing operations must ensure appropriate storage conditions are maintained and allow for appropriate security of stocks.

Stock should be rotated according to the 'first expiry, first out' (FEFO) principle. Exceptions should be documented.

Medicinal products should be handled and stored in such a manner as to prevent spillage, breakage, contamination and mix-ups. Medicinal products should not be stored directly on the floor unless the package is designed to allow such storage (such as for some medicinal gas cylinders).

Medicinal products that are nearing their expiry date/shelf life should be withdrawn immediately from saleable stock either physically or through other equivalent electronic segregation.

Stock inventories should be performed regularly taking into account national legislation requirements. Stock irregularities should be investigated and documented.

5.6 Destruction of obsolete goods

Medicinal products intended for destruction should be appropriately identified, held separately and handled in accordance with a written procedure.

Destruction of medicinal products should be in accordance with national or international requirements for handling, transport and disposal of such products.

Records of all destroyed medicinal products should be retained for a defined period.

5.7 Picking

Controls should be in place to ensure the correct product is picked. The product should have an appropriate remaining shelf life when it is picked.

5.8 Supply

For all supplies, a document (e.g. delivery note) must be enclosed stating the date; name and pharmaceutical form of the medicinal product, batch number at least for products bearing the safety features; quantity supplied; name and address of the supplier, name and delivery address of the consignee[1] (actual physical storage premises, if different) and applicable transport and storage conditions. Records should be kept so that the actual location of the product can be known.

[1] Article 82 of Directive 2001/83/EC.

5.9 Export to third countries

The export of medicinal products falls within the definition of 'wholesale distribution'[2]. A person exporting medicinal products must hold a wholesale distribution authorisation or a manufacturing authorisation. This is also the case if the exporting wholesale distributor is operating from a free zone.

The rules for wholesale distribution apply in their entirety in the case of export of medicinal products. However, where medicinal products are exported, they do not need to be covered by a marketing authorisation of the Union or a Member State[3]. Wholesalers should take the appropriate measures in order to prevent these medicinal products reaching the Union market. Where wholesale distributors supply medicinal products to persons in third countries, they shall ensure that such supplies are only made to persons who are authorised or entitled to receive medicinal products for wholesale distribution or supply to the public in accordance with the applicable legal and administrative provisions of the country concerned.

Chapter 6 – Complaints, Returns, Suspected Falsified Medicinal Products and Medicinal Product Recalls

6.1 Principle

All complaints, returns, suspected falsified medicinal products and recalls must be recorded and handled carefully according to written procedures. Records should be made available to the competent authorities. An assessment of returned medicinal products should be performed before any approval for resale. A consistent approach by all partners in the supply chain is required in order to be successful in the fight against falsified medicinal products.

6.2 Complaints

Complaints should be recorded with all the original details. A distinction should be made between complaints related to the quality of a medicinal product and those related to distribution. In the event of a complaint about the quality of a medicinal product and a potential product defect, the manufacturer and/or marketing authorisation holder should be informed without delay. Any product distribution complaint should be thoroughly investigated to identify the origin of or reason for the complaint.

[2] Article 1(17) of Directive 2001/83/EC.
[3] Article 85(a) of Directive 2001/83/EC.

A person should be appointed to handle complaints and allocated sufficient support personnel.

If necessary, appropriate follow-up actions (including CAPA) should be taken after investigation and evaluation of the complaint, including where required notification to the national competent authorities.

6.3 Returned medicinal products

Returned products must be handled according to a written, risk-based process taking into account the product concerned, any specific storage requirements and the time elapsed since the medicinal product was originally dispatched. Returns should be conducted in accordance with national law and contractual arrangements between the parties.

Medicinal products which have left the premises of the distributor should only be returned to saleable stock if all of the following are confirmed:

(i) the medicinal products are in their unopened and undamaged secondary packaging and are in good condition; have not expired and have not been recalled;
(ii) medicinal products returned from a customer not holding a wholesale distribution authorisation or from pharmacies authorised to supply medicinal products to the public should only be returned to saleable stock if they are returned within an acceptable time limit, for example 10 days;
(iii) it has been demonstrated by the customer that the medicinal products have been transported, stored and handled in compliance with their specific storage requirements;
(iv) they have been examined and assessed by a sufficiently trained and competent person authorised to do so;
(v) the distributor has reasonable evidence that the product was supplied to that customer (via copies of the original delivery note or by referencing invoice numbers, etc.) and the batch number for products bearing the safety features is known, and that there is no reason to believe that the product has been falsified.

Moreover, for medicinal products requiring specific temperature storage conditions such as low temperature, returns to saleable stock can only be made if there is documented evidence that the product has been stored under the authorised storage conditions throughout the entire time. If any deviation has occurred a risk assessment has to be performed, on which basis the integrity of the product can be demonstrated. The evidence should cover:

(i) delivery to customer;
(ii) examination of the product;

(iii) opening of the transport packaging;
(iv) return of the product to the packaging;
(v) collection and return to the distributor;
(vi) return to the distribution site refrigerator.

Products returned to saleable stock should be placed such that the 'first expired first out' (FEFO) system operates effectively.

Stolen products that have been recovered cannot be returned to saleable stock and sold to customers.

6.4 Falsified medicinal products

Wholesale distributors must immediately inform the competent authority and the marketing authorisation holder of any medicinal products they identify as falsified or suspect to be falsified[1]. A procedure should be in place to this effect. It should be recorded with all the original details and investigated.

Any falsified medicinal products found in the supply chain should immediately be physically segregated and stored in a dedicated area away from all other medicinal products. All relevant activities in relation to such products should be documented and records retained.

6.5 Medicinal product recalls

The effectiveness of the arrangements for product recall should be evaluated regularly (at least annually).

Recall operations should be capable of being initiated promptly and at any time.

The distributor must follow the instructions of a recall message, which should be approved, if required, by the competent authorities.

Any recall operation should be recorded at the time it is carried out. Records should be made readily available to the competent authorities.

The distribution records should be readily accessible to the person(s) responsible for the recall, and should contain sufficient information on distributors and directly supplied customers (with addresses, phone and/or fax numbers inside and outside working hours, batch numbers at least for medicinal products bearing safety features as required by legislation and quantities delivered), including those for exported products and medicinal product samples.

The progress of the recall process should be recorded for a final report.

[1] Article 80(i) of Directive 2001/83/EC.

Chapter 7 – Outsourced Activities

7.1 Principle

Any activity covered by the GDP guide that is outsourced should be correctly defined, agreed and controlled in order to avoid misunderstandings which could affect the integrity of the product. There must be a written contract between the contract giver and the contract acceptor which clearly establishes the duties of each party.

7.2 Contract giver

The contract giver is responsible for the activities contracted out.

The contract giver is responsible for assessing the competence of the contract acceptor to successfully carry out the work required and for ensuring by means of the contract and through audits that the principles and guidelines of GDP are followed. An audit of the contract acceptor should be performed before commencement of, and whenever there has been a change to, the outsourced activities. The frequency of audit should be defined based on risk depending on the nature of the outsourced activities. Audits should be permitted at any time.

The contract giver should provide the contract acceptor with all the information necessary to carry out the contracted operations in accordance with the specific product requirements and any other relevant requirements.

7.3 Contract acceptor

The contract acceptor should have adequate premises and equipment, procedures, knowledge and experience, and competent personnel to carry out the work ordered by the contract giver.

The contract acceptor should not pass to a third party any of the work entrusted to him under the contract without the contract giver's prior evaluation and approval of the arrangements and an audit of the third party by the contract giver or the contract acceptor. Arrangements made between the contract acceptor and any third party should ensure that the wholesale distribution information is made available in the same way as between the original contract giver and contract acceptor.

The contract acceptor should refrain from any activity which may adversely affect the quality of the product(s) handled for the contract giver.

The contract acceptor must forward any information that can influence the quality of the product(s) to the contract giver in accordance with the requirement of the contract.

Chapter 8 – Self-inspections

8.1 Principle

Self-inspections should be conducted in order to monitor implementation and compliance with GDP principles and to propose necessary corrective measures.

8.2 Self-inspections

A self-inspection programme should be implemented covering all aspects of GDP and compliance with the regulations, guidelines and procedures within a defined time frame. Self-inspections may be divided into several individual self-inspections of limited scope.

Self-inspections should be conducted in an impartial and detailed way by designated competent company personnel. Audits by independent external experts may also be useful but may not be used as a substitute for self-inspection.

All self-inspections should be recorded. Reports should contain all the observations made during the inspection. A copy of the report should be provided to the management and other relevant persons. In the event that irregularities and/or deficiencies are observed, their cause should be determined and the corrective and preventive actions (CAPA) should be documented and followed up.

Chapter 9 – Transportation

9.1 Principle

It is the responsibility of the supplying wholesale distributor to protect medicinal products against breakage, adulteration and theft, and to ensure that temperature conditions are maintained within acceptable limits during transport.

Regardless of the mode of transport, it should be possible to demonstrate that the medicines have not been exposed to conditions that may compromise their quality and integrity. A risk-based approach should be utilised when planning transportation.

9.2 Transportation

The required storage conditions for medicinal products should be maintained during transportation within the defined limits as described by the manufacturers or on the outer packaging.

If a deviation such as temperature excursion or product damage has occurred during transportation, this should be reported to the distributor and recipient of the affected medicinal products. A procedure should also be in place for investigating and handling temperature excursions.

It is the responsibility of the wholesale distributor to ensure that vehicles and equipment used to distribute, store or handle medicinal products are suitable for their use and appropriately equipped to prevent exposure of the products to conditions that could affect their quality and packaging integrity.

There should be written procedures in place for the operation and maintenance of all vehicles and equipment involved in the distribution process, including cleaning and safety precautions.

Risk assessment of delivery routes should be used to determine where temperature controls are required. Equipment used for temperature monitoring during transport within vehicles and/or containers, should be maintained and calibrated at regular intervals at least once a year.

Dedicated vehicles and equipment should be used, where possible, when handling medicinal products. Where non-dedicated vehicles and equipment are used procedures should be in place to ensure that the quality of the medicinal product will not be compromised.

Deliveries should be made to the address stated on the delivery note and into the care or the premises of the consignee. Medicinal products should not be left on alternative premises.

For emergency deliveries outside normal business hours, persons should be designated and written procedures should be available.

Where transportation is performed by a third party, the contract in place should encompass the requirements of Chapter 7. Transportation providers should be made aware by the wholesale distributor of the relevant transport conditions applicable to the consignment. Where the transportation route includes unloading and reloading or transit storage at a transportation hub, particular attention should be paid to temperature monitoring, cleanliness and the security of any intermediate storage facilities.

Provision should be made to minimise the duration of temporary storage while awaiting the next stage of the transportation route.

9.3 Containers, packaging and labelling

Medicinal products should be transported in containers that have no adverse effect on the quality of the products, and that offer adequate protection from external influences, including contamination.

Selection of a container and packaging should be based on the storage and transportation requirements of the medicinal products; the space required for the amount of medicines; the anticipated external temperature extremes; the estimated maximum time for transportation including transit storage at customs; the qualification status of the packaging and the validation status of the shipping containers.

Containers should bear labels providing sufficient information on handling and storage requirements and precautions to ensure that the products are properly handled and secured at all times. The containers should enable identification of the contents of the containers and the source.

9.4 Products requiring special conditions

In relation to deliveries containing medicinal products requiring special conditions such as narcotics or psychotropic substances, the wholesale distributor should maintain a safe and secure supply chain for these products in accordance with requirements laid down by the Member States concerned. There should be additional control systems in place for delivery of these products. There should be a protocol to address the occurrence of any theft.

Medicinal products comprising highly active and radioactive materials should be transported in safe, dedicated and secure containers and vehicles. The relevant safety measures should be in accordance with international agreements and national legislation.

For temperature-sensitive products, qualified equipment (e.g. thermal packaging, temperature-controlled containers or temperature controlled vehicles) should be used to ensure correct transport conditions are maintained between the manufacturer, wholesale distributor and customer.

If temperature-controlled vehicles are used, the temperature monitoring equipment used during transport should be maintained and calibrated at regular intervals. Temperature mapping under representative conditions should be carried out and should take into account seasonal variations.

If requested, customers should be provided with information to demonstrate that products have complied with the temperature storage conditions.

If cool packs are used in insulated boxes, they need to be located such that the product does not come in direct contact with the cool pack. Staff must be trained on the procedures for assembly of the insulated boxes (seasonal configurations) and on the re-use of cool packs.

There should be a system in place to control the re-use of cool packs to ensure that incompletely cooled packs are not used in error. There should be adequate physical segregation between frozen and chilled ice packs.

The process for delivery of sensitive products and control of seasonal temperature variations should be described in a written procedure.

Chapter 10 – Specific Provisions for Brokers[1]

10.1 Principle

A 'broker' is a person involved in activities in relation to the sale or purchase of medicinal products, except for wholesale distribution, that do not include physical handling and that consist of negotiating independently and on behalf of another legal or natural person.[2]

Brokers are subject to a registration requirement. They must have a permanent address and contact details in the Member State where they are registered[3]. They must notify the competent authority of any changes to those details without unnecessary delay.

By definition, brokers do not procure, supply or hold medicines. Therefore, requirements for premises, installations and equipment as set out in Directive 2001/83/EC do not apply. However, all other rules in Directive 2001/83/EC that apply to wholesale distributors also apply to brokers.

10.2 Quality system

The quality system of a broker should be defined in writing, approved and kept up to date. It should set out responsibilities, processes and risk management in relation to their activities.

The quality system should include an emergency plan which ensures effective recall of medicinal products from the market ordered by the manufacturer or the competent authorities or carried out in cooperation with the manufacturer or marketing authorisation holder for the medicinal product concerned[4]. The competent authorities must be immediately informed of any suspected falsified medicines offered in the supply chain[5].

[1] Article 85b(3) of Directive 2001/83/EC.
[2] Article 1(17a) of Directive 2001/83/EC.
[3] Article 85b of Directive 2001/83/EC.
[4] Article 80(d) of Directive 2001/83/EC.
[5] Article 85b(1), third paragraph of Directive 2001/83/EC.

10.3 Personnel

Any member of personnel involved in the brokering activities should be trained in the applicable EU and national legislation and in the issues concerning falsified medicinal products.

10.4 Documentation

The general provisions on documentation in Chapter 4 apply.

In addition, at least the following procedures and instructions, along with the corresponding records of execution, should be in place:

(i) procedure for complaints handling;
(ii) procedure for informing competent authorities and marketing authorisation holders of suspected falsified medicinal products;
(iii) procedure for supporting recalls;
(iv) procedure for ensuring that medicinal products brokered have a marketing authorisation;
(v) procedure for verifying that their supplying wholesale distributors hold a distribution authorisation, their supplying manufacturers or importers hold a manufacturing authorisation and their customers are authorised to supply medicinal products in the Member State concerned;
(vi) records should be kept either in the form of purchase/sales invoices or on computer, or in any other form for any transaction in medicinal products brokered and should contain at least the following information: date; name of the medicinal product; quantity brokered; name and address of the supplier and the customer; and batch number at least for products bearing the safety features.

Records should be made available to the competent authorities, for inspection purposes, for the period stated in national legislation but at least five years.

Chapter 11 – Final Provisions

These Guidelines replace the Guidelines on Good Distribution Practice of medicinal products for human use, published on 1 March 1994[1] and the Guidelines of 7 March 2013 on Good Distribution Practice of medicinal products for human use[2].

These Guidelines will be applied from the first day following their publication in the *Official Journal of the European Union*. EN C 343/12 Official Journal of the European Union 23.11.2013.

[1] OJ C 63, 1.3.1994, p. 4.
[2] OJ C 68, 8.3.2013, p. 1.

Annex

Glossary of terms

Term	Definition
Good Distribution Practice (GDP)	GDP is that part of quality assurance which ensures that the quality of medicinal products is maintained throughout all stages of the supply chain from the site of manufacturer to the pharmacy or person authorised or entitled to supply medicinal products to the public.
Export procedure	Export procedure: allow Community goods to leave the customs territory of the Union. For the purpose of these guidelines, the supply of medicines from EU Member State to a contracting State of the European Economic Area is not considered as export.
Falsified medicinal product[1]	Any medicinal product with a false representation of: (a) its identity, including its packaging and labelling, its name or its composition as regards any of the ingredients including excipients and the strength of those ingredients; (b) its source, including its manufacturer, its country of manufacturing, its country of origin or its marketing authorisation holder; or (c) its history, including the records and documents relating to the distribution channels used.
Free zones and free warehouses[2]	Free zones and free warehouses are parts of the customs territory of the Community or premises situated in that territory and separated from the rest of it in which: (a) Community goods are considered, for the purpose of import duties and commercial policy import measures, as not being on Community customs territory, provided they are not released for free circulation or placed under another customs procedure or used or consumed under conditions other than those provided for in customs regulations; (b) Community goods for which such provision is made under Community legislation governing specific fields qualify, by virtue of being placed in a free zone or free warehouse, for measures normally attaching to the export of goods.

(*Continued*)

Glossary of terms (*Continued*)

Term	Definition
Holding	Storing medicinal products.
Transport	Moving medicinal products between two locations without storing them for unjustified periods of time.
Procuring	Obtaining, acquiring, purchasing or buying medicinal products from manufacturers, importers or other wholesale distributors.
Qualification	Action of proving that any equipment works correctly and actually leads to the expected results. The word 'validation' is sometimes widened to incorporate the concept of qualification. (Defined in EudraLex Volume 4 Glossary to the GMP Guidelines).
Supplying	All activities of providing, selling, donating medicinal products to wholesalers, pharmacists, or persons authorised or entitled to supply medicinal products to the public.
Quality risk management	A systematic process for the assessment, control, communication and review of risks to the quality of the drug (medicinal) product across the product lifecycle.
Quality system	The sum of all aspects of a system that implements quality policy and ensures that quality objectives are met. (International Conference on Harmonisation of Technical Requirements for Registration of Pharmaceuticals for Human Use, Q9).
Validation	Action of proving that any procedure, process, equipment, material, activity or system actually leads to the expected results (see also 'qualification'). (Defined in EudraLex Volume 4 Glossary to the GMP Guidelines)

[1]Article 1(33) of Directive 2001/83/EC.
[2]Articles 166 to Article 181 of Council Regulation (EEC) No 2913/92 of 12 October 1992 establishing the Community Customs Code, (OJ L 302, 19.10.1992, p. 1).

EUROPEAN COMMISSION Q&A ON GDP GUIDELINES

In March 2014 the European Commission published a question and answer document responding to frequently asked questions in relation to the guidelines on Good Distribution Practice of medicinal products for human use. This document sets out frequently asked 'questions and answers' regarding the new guidelines on Good Distribution Practice of medicinal products for human use applicable as of 8 September 2013,[1] and their revision of November 2013[2].

It should be noted that the views expressed in this questions and answers document are not legally binding. Ultimately, only the European Court of Justice can give an authoritative interpretation of Union law.

1. QUESTION: IN CHAPTER 2 – PERSONNEL, 2.3.(1), WHAT IS THE DEFINITION OF "COMPETENT"?

Answer: Having the necessary experience and/or training to adequately perform the job.

2. QUESTION: CONCERNING CHAPTER 2 – PERSONNEL, 2.4.(1), IS IT POSSIBLE TO EMPLOY STAFF STARTING THEIR PROFESSIONAL CAREER?

Answer: Yes, if you properly training them before assigning them to their tasks.

3. QUESTION: IN CHAPTER 2 – PERSONNEL, 2.4.(5), WHAT IS MEANT BY "PERIODICALLY ASSESSED"?

Answer: Assessed at regular intervals. The timing of intervals is left flexible, without prejudice to national legislation.

4. QUESTION: IN CHAPTER 2 – PERSONNEL, 2.5, DOES THE STATEMENT "APPROPRIATE PROCEDURES RELATING TO PERSONNEL HYGIENE, RELEVANT TO THE ACTIVITIES BEING CARRIED OUT, SHOULD BE ESTABLISHED AND OBSERVED" REFER TO THE HEALTH AND/OR CLEANLINESS OF THE STAFF?

Answer: It only refers to the cleanliness of the staff, so to avoid any alteration of the product.

5. QUESTION: IN CHAPTER 3 – PREMISES AND EQUIPMENT, 3.2.(3), IS THE INTENT OF SEGREGATION TO AVOID "CROSS-CONTAMINATION" AS MENTIONED IN CHAPTER 5?

---

[1] Guidelines of 7 March 2013 on Good Distribution Practice of Medicinal Products for Human Use, OJ C 68, 8.3.2013, p. 1.

[2] Guidelines of 5 November 2013 on Good Distribution Practice of Medicinal Products for Human Use, OJ C 343, 23.11.2013, p. 1.

Answer: The intent of this provision is to avoid handling errors and accidental swaps of products. This is why electronic segregation is allowed, except for falsified, expired, recalled and rejected products which always have to be segregated physically.

6. QUESTION: CONCERNING CHAPTER 3 – PREMISES AND-EQUIPMENT, 3.2.(7), IS THE LID OF A BOX / TOTE AN ADEQUATE PROTECTION AGAINST WEATHER CONDITIONS?

Answer: Any tool protecting products from prevailing weather conditions is acceptable.

7. QUESTION: CONCERNING CHAPTER 3 – PREMISES AND EQUIPMENT, 3.2.1, HOW MANY PROBES ARE NECESSARY TO MONITOR THE TEMPERATURE?

Answer: The number of probes and their placement depend on the risk analysis performed on the site and the placement should be in agreement with the mapping results.

8. QUESTION: CONCERNING CHAPTER 3 – PREMISES AND EQUIPMENT, 3.2.1.(2), WHAT IS MEANT BY INITIAL TEMPERATURE MAPPING?

Answer: An initial temperature mapping is an exercise in which temperature sensors are placed on the points identified as most critical through a risk analysis (e.g. at different heights, near a sunny window, next to the doors, etc.). Once placed, a measurement is taken over a period of time and with the results obtained, the temperature sensors will be places where greater fluctuation occurred. The mapping should be performed in different seasons where highest and lowest temperatures are reached.

9. QUESTION: CONCERNING CHAPTER 3 – PREMISES AND EQUIPMENT, 3.3.(3), WHAT ARE APPROPRIATE SETTINGS FOR THE ALARM LEVELS?

Answer: Alarm settings should be chosen as to guarantee a timely alert of personnel when there are excursions from predefined storage conditions

10. QUESTION: CONCERNING CHAPTER 3 – PREMISES AND EQUIPMENT, 3.3.2.(1), IS IT REQUIRED TO HAVE ONLY EQUIPMENT WITH A CE CERTIFICATE OF CONFORMITY?

Answer: No. The CE marking is mandatory only for products dating from 1993 or later. It should be noted that the presence of a CE certificate of Conformity doesn't exempt from equipment validation/qualification.

11. QUESTION: CONCERNING CHAPTER 3 – PREMISES AND EQUIPMENT, 3.3.2.(1), IS IT REQUIRED TO RECORD ALL DEVIATIONS OR CAN THEY BE LIMITED TO SIGNIFICANT

DEVIATIONS HAVING AN IMPACT ON PRODUCT SECURITY AND INTEGRITY?

Answer: All deviations from established procedures should be documented.

12. QUESTION: CONCERNING CHAPTER 5 – OPERATIONS, 5.3.(3), ARE THERE ANY BEST PRACTICES FOR THE INVESTIGATION OF "UNUSUAL SALES PATTERNS"?

Answer: It is recommended to check for unusual repetition of orders, sudden increases of orders, and unusually low prices.

13. QUESTION: CONCERNING CHAPTER 5 – OPERATIONS, 5.4.(3), HOW CAN WHOLESALERS PROVE THAT THE MEDICINES THEY RECEIVE ARE RELEASED FOR SALE?

Answer: The batches of medicinal products which have undergone the controls referred to in Art. 51 of Directive 2001/83/EC in a Member State are exempt from the controls if they are marketed in another Member State, accompanied by the control reports signed by the qualified person. In other words, the batch arriving from another member state needs to be accompanied by evidence that the manufacturer's qualified person has certified the finished product batch for the target member state. The technical means by which this evidence is provided has been left to the discretion of the companies: a physical control report or an equivalent system of proof of release.

14. QUESTION: CONCERNING CHAPTER 5 – OPERATIONS, 5.4.(3), CAN THE PROVISION REQUIRING WHOLESALERS TO PROVE THAT THE MEDICINES THEY RECEIVE ARE RELEASED FOR SALE BE MADE APPLICABLE AS OF 2017, WHEN MEDICINAL PRODUCTS WILL CARRY SAFETY FEATURES THAT CAN BE CHECKED THROUGH A DATABASE?

Answer: No, these provisions are already into operation as the new GDP guideline simply repeats what mentioned in Directive 2001/83/EC.

15. QUESTION: IN CHAPTER 5 – OPERATIONS, 5.5.(1), WHAT IS THE DEFINITION OF "SEPARATELY"?

Answer: This provision refers to physical separation only in case of presence of other products likely to alter the medicinal product. Segregation is also necessary for falsified, expired, recalled and rejected medicines. In all other cases, electronic segregation is possible.

16. QUESTION: CONCERNING CHAPTER 5 – OPERATIONS, 5.5.(8), IN WHAT WAY SHOULD INVESTIGATION AND DOCUMENTATION BE DONE?

Answer: Reasons for stock irregularities should be investigated and documented.

17. QUESTION: CONCERNING CHAPTER 6 – COMPLAINTS, RETURNS, SUSPECTED FALSIFIED MEDICINAL PRODUCTS AND MEDICINAL PRODUCT RECALLS, 6.3.(2)(I), HOW CAN WHOLESALERS VERIFY THAT A PACK HAS NOT BEEN OPENED IF IT DOES NOT CARRY AN ANTI-TAMPER DEVICE?

Answer: The anti-tampering device will become compulsory in the coming years. Meanwhile, check to the best of your capacities whether the pack has signs of tampering.

18. QUESTION: CONCERNING CHAPTER 6 – COMPLAINTS, RETURNS, SUSPECTED FALSIFIED MEDICINAL PRODUCTS AND MEDICINAL PRODUCT RECALLS, 6.3.(2)(III), HOW SHOULD THE CUSTOMER DEMONSTRATE THAT THE MEDICINAL PRODUCTS HAVE BEEN TRANSPORTED, STORED AND HANDLED IN COMPLIANCE WITH SPECIFIC STORAGE REQUIREMENTS?

Answer: The customer needs to provide papers showing that the medicinal products have been transported, stored and handled in compliance with their specific storage requirements.

19. QUESTION: CONCERNING CHAPTER 6 – COMPLAINTS, RETURNS, SUSPECTED FALSIFIED MEDICINAL PRODUCTS AND MEDICINAL PRODUCT RECALLS, 6.3.(3), WHAT SHOULD THE DOCUMENTED EVIDENCE CONTAIN TO BE REGARDED AS SUFFICIENT?

Answer: The evidence should cover:

(i) delivery to customer;
(ii) examination of the product;
(iii) opening of the transport packaging;
(iv) return of the product to the packaging;
(v) collection and return to the distributor;
(vi) return to the distribution site refrigerator.

20. QUESTION: IN CHAPTER 6 – COMPLAINTS, RETURNS, SUSPECTED FALSIFIED MEDICINAL PRODUCTS AND MEDICINAL PRODUCT RECALLS, 6.3.(3)(IV), DOES THE TERM "PACKAGING" REFER TO THE TRANSPORT BOX?

Answer: Yes

21. QUESTION: IN CHAPTER 6 – COMPLAINTS, RETURNS, SUSPECTED FALSIFIED MEDICINAL PRODUCTS AND MEDICINAL PRODUCT RECALLS, 6.5.(6), WHAT DOES IT MEAN TO RECORD THE PROGRESS OF THE RECALL?

Answer: You need to record the status of the recall and whether it was successful or not.

22. QUESTION: CONCERNING CHAPTER 9 – TRANSPORTATION, 9.2.(1), CAN WE DEVIATE FROM STORAGE CONDITIONS IF THE MANUFACTURER AGREES TO THE TRANSPORTATION OF THE PRODUCT WITHIN A CERTAIN TEMPERATURE RANGE (2°–25°C) FOR A LIMITED TIME FRAME OF 6 HOURS?

Answer: No. Storage temperature limits as described by the manufacturer or on the outer packaging need to be respected for each stage of transport during the whole transport chain.

23. QUESTION: CONCERNING CHAPTER 9 – TRANSPORTATION, 9.2.(9), ARE TRANSPORTATION COMPANIES REQUIRED TO HOLD A WHOLESALE DISTRIBUTION AUTHORISATION AND COMPLY WITH GDP?

Answer: Transport companies do not need to hold a wholesale distribution authorisation to transport medicinal products. However, they should follow the parts of the GDP guideline relevant to their activities, amongst others Chapter 9.

24. QUESTION: IN CHAPTER 9 – TRANSPORTATION, 9.4.(2), DOES THE DEFINITION OF HIGHLY ACTIVE MATERIAL COMPRISE CYTOSTATICS?

Answer: The guidelines do not include a list of highly active materials, or a definition of highly active material. Without prejudice to specific provisions included in national legislation, the wording leaves a margin of manoeuvre to the manufacturer/distributor.

25. QUESTION: IN CHAPTER 9 – TRANSPORTATION 9.4.(7), WHAT IS THE DEFINITION OF "CHILLED"?

Answer: "Chilled" refers to refrigerated storage conditions.

CHAPTER

5

UK Guidance on Wholesale Distribution Practice

Contents

The Application and Inspection Process: "What to Expect" 114
Application 114
Inspection 115
Risk-based Inspection Programme 118
Introduction 118
Sentinel risk information module 119
Current implementation status 120
GDP risk-based inspection (RBI) programme 120
Compliance report 120
Risk rating process 120
Conditions of Holding a Wholesale Dealer's Licence 121
Appointment and Duties of the Responsible Person 128
The Responsible Person Gold Standard 130
Quality Management 135
Introduction 135
Quality system 135
Quality Risk Management (QRM) 135
Change control 135
Deviation management and Corrective and Preventative Actions (CAPA) 136
Management review and monitoring 136
Controls on Certain Medicinal Products 137
Controlled drugs and precursor chemicals 138
Use of the mandatory requisition form for Schedule 2 and 3 controlled drugs 139
Controls on strategic goods and drugs useable in execution by lethal injection 141
Control of lisdexamfetamine, tramadol, zaleplon, zopiclone and reclassification of ketamine 142
Controls and authorisations applying to those handling medicinal products derived from blood 144
Procurement, storage, distribution and disposal of vaccines 144
Wholesale of radiopharmaceuticals 144
Wholesale of veterinary medicinal products 144
Sourcing and Exporting Medicinal Products – Non-EEA Countries 145
Introduced medicinal products 146
Export to a non-EEA country 148
Qualification check in non-EEA countries 150
Wholesale obligations for exported and introduced medicinal products 151
Incoterms® 151
Import of medicinal products 153
Ensuring you have the correct authorisations for the intended activity 154
Temperature Control and Monitoring 155
Temperature mapping 156
Refrigerated and ambient medicinal products, receipt, storage, packing and transportation 158

Control and monitoring of storage areas 160
Best practice 160
Calibration of temperature monitoring devices including ambient 161
Packing of consignments and temperature management during transportation 162
Handling Returns 164
Returns of refrigerated medicinal products 164
Returns of non-defective medicinal products 164
Short-Term Storage of Ambient and Refrigerated Medicinal Products – Requirements for a Wholesale Dealer's Licence 166
Sales Representative Samples 167
Qualification of Customers and Suppliers 168
Qualification of suppliers 168
Customer qualification 170
Falsified Medicines 171
Regulatory Action 173
Diverted Medicines 173
Parallel Importation 173
Parallel Distribution 175
Continued Supply 175
Matters Relating to Unlicensed Medicines 177
Importation of unlicensed medicines and centrally authorised products 178
Supply of unlicensed medicines when an equivalent licensed product becomes available 179
Reporting Adverse Reactions 180
Product Recall/Withdrawal 180
Issue a "Recall" 181
Issue a "Drug Alert" 181
Management of the recall 181
Management of recall activity by wholesalers 182
Testing the recall process 183
Follow-up action 184
Reporting a suspected defect 185
Data Integrity 185
Basic concepts 185
Criticality of data 186
ALCOA applied to GDP 187
Data Governance 189

The Application and Inspection Process: "What to Expect"

Application

To sell or supply medicines to anyone other than the patient using the medicine, you need a wholesale dealer's licence – also known as a wholesale distribution authorisation or WDA(H).

Applicants for a new wholesale dealer's licence or existing licence holders wishing to vary their licence should follow MHRA guidance and instructions and make their application using the MHRA Process Licensing Portal accessible via the GOV.UK website.[1]

MHRA acting as the licensing authority will only issue a wholesale dealer's licence when it is satisfied, following an inspection of the site(s), that the information contained in the application is accurate and in compliance with the requirements of the legislation.

[1] https://www.gov.uk/guidance/apply-for-manufacturer-or-wholesaler-of-medicines-licences.

When appropriate, MHRA may refuse to grant a wholesale dealer's licence or may grant a wholesale dealer's licence otherwise than as applied for. In such cases the licensing authority will notify the applicant of its proposals. The notification will set out the reasons for its proposals and give the applicant a period of not less than 28 days to respond.

APPLY FOR A LICENCE FOR HUMAN OR VETERINARY MEDICINES

The licence applicant must provide the following:

- A separate application for a human medicines wholesale dealer's licence and a veterinary medicines wholesale dealer's licence.
- An application fee (which includes the fee relating to the first inspection).
- Company and site details.
- Nomination of one or more people to be named as Responsible Person (RP) for a human medicines wholesale dealer's licence or a wholesale dealer's Qualified Person on a veterinary medicines wholesale dealer's licence. This should be supported with CV, references and identification details for each named person.
- Details of the activities to be carried out and the medicinal products to be wholesaled. For all activities and products included in the application there must be sufficient provisions on site to support the application.

The licence applicant should also consider if any of the following are required; separate applications are required for each:

- Broker registration.
- Active Substance registration.
- Registration to sell medicines on the internet ("common logo").

APPLY TO VARY A LICENCE

The licence holder is required to keep their licence up to date. Any changes to named personnel, sites, activities or categories of products wholesaled require a licence variation. This variation may trigger an inspection which must take place before the variation can be approved. Wholesale from a new site, new activities or new products may not commence until the replacement licence is received.

Inspection

PLANNING

Fee bearing inspections of licensed wholesale dealers are carried out to assess the degree of compliance to standards of Good Distribution Practice (GDP) and compliance with the provisions of the licence.

Inspections of licensed wholesale dealers are undertaken as part of the risk-based inspection programme, further details of which can be found later in this guide.

NOTIFICATION

Advance notice of inspection is normally given to a company, unless circumstances require that an unannounced inspection should take place. The timing of the inspection would normally be notified in writing by the inspector.

In accordance with the GDP risk-based inspection process, sites will be required to complete a Compliance Report in advance of inspection. Further information and guidance can be found in the risk-based inspections section.

CONDUCT

The major stages of the inspection process are:

- the introductory or opening meeting;
- the detailed site inspection;
- the summary or closing meeting.

INTRODUCTORY OR OPENING MEETING

The purpose of the meeting is for the inspector to meet with the appropriate key personnel from the company to discuss the arrangements for the inspection. The inspector would typically confirm the nature of the business, premises and security arrangements, areas to be visited and any documentation which may be required.

SITE INSPECTION

The purpose of the site inspection is to determine the degree of conformity of the operations to the requirements of GDP and to assess compliance with the terms and conditions of licences issued under the appropriate legislation or with details submitted in support of an application for a licence. The inspection will typically involve visits to goods receipt, storage and dispatch areas (including ambient and refrigerated), returns/quarantine area, interviews with key personnel and a review of stock movement and quality system documentation including product recalls. Any observations, recommendations and deficiencies noted during the inspection would normally be discussed with the company representatives at the time.

During inspections of manufacturing and wholesale operations, samples of starting materials, work in progress and finished products may be taken for testing if an inspector considers that this might assist in the detection of quality deficiencies. Occasionally, samples may be taken when these cannot be obtained from other sources, for routine surveillance purposes.

The inspection will also assess if the named RP meets the knowledge, experience and responsibility requirements. This will be more noticeable when a person has been newly nominated or if their role has changed substantially.

The licence applicant/licence holder must be prepared to demonstrate that all proposed sites, activities, products and personnel are suitable, within the scope of the licence and accurately reflected on the licence. A table is provided in Appendix 4 to assist the licence holder/applicant in ensuring they have the correct licence when planning to import or export.

SUMMARY OR CLOSING MEETING

The purpose of the meeting is for the inspector to provide the company with a verbal summary of the inspection findings and to allow the company to correct at this stage any misconceptions. The inspector would typically summarise the definition and classification of deficiencies they propose to report and the company are encouraged to give an undertaking to resolve the deficiencies and to agree a provisional timetable for corrective action. The inspector would also describe the arrangements for the formal notification of the deficiencies to the company (the post-inspection letter) and what is expected as a response.

Deficiencies are classified as follows:

- *Critical deficiency:*

  Any departure from Guidelines on GDP resulting in a medicinal product causing a significant risk to the patient and public health. This includes an activity increasing the risk of falsified medicines reaching the patients. A combination of a number of major deficiencies that indicates a serious systems failure. An example of a critical deficiency could be:

  Purchase from or supply of medicinal products to a non-authorised person; Storage of products requiring refrigeration at ambient temperatures; Rejected or recalled products found in sellable stock.

- *Major deficiency:*

  A non-critical deficiency:

  - which indicates a major deviation from GDP; **or**
  - which has caused or may cause a medicinal product not to comply with its marketing authorisation in particular its storage and transport conditions; **or**
  - which indicates a major deviation from the terms and provisions of the wholesale distribution authorisation; **or** a combination of several other deficiencies, none of which on their own may be major, but which may together represent a major deficiency.

- *Other deficiency*:

  A deficiency which cannot be classified as either critical or major, but which indicates a departure from guidelines on GDP.

  The choice of company representatives at the meeting is primarily for the company to decide, but should normally include the senior staff who were present during the inspection and the RP.

  Depending upon the inspection findings and the response from the company during and following the inspection, the inspector may take one of a number of actions ranging from:

  – issuing a GDP certificate confirming essential compliance with GDP; to
  – referral to the compliance escalation process or the Inspection Action Group (IAG) for consideration for adverse licensing action where serious non-compliance is found.

Further information on the Compliance escalation process and IAG can be found in this publication.

COMPANY RESPONSES

The inspected site is expected to provide a written response (by letter or email) to the post-inspection letter within the required timeframe. The response should consider the context of the deficiency within the overall quality system rather than just the specific issue identified. The response should include proposals for dealing with the deficiencies, together with a timetable for their implementation. The response should be structured as follows:

- Restate the deficiency number and the deficiency as written below.
- State the proposed corrective action.
- State the proposed target date for the completion of the corrective action(s).
- Include any comment the company considers appropriate.
- Provide evidence supporting any corrective action where it is considered appropriate.

INSPECTION REPORT

Once the inspector is satisfied that any necessary remedial action has been taken or is in hand and that the site is essentially in compliance with GDP, an inspection report and GDP certificate are finalised.

Risk-based Inspection Programme

Introduction

MHRA has been incorporating elements of risk management into its inspection programme for a number of years. A formal risk-based inspection (RBI) programme was implemented on 1 April 2009, following public consultation MLX 345. The RBI programme covers all aspects of good practices associated with the inspection of clinical, pre-clinical and quality control laboratories, clinical trials, manufacturers, wholesalers and pharmacovigilance systems. The primary aim of the RBI programme is to enable inspectorate resources to focus on areas that maximise protection of public health while reducing the overall administrative and economic burden to stakeholders.

Sentinel risk information module

Working with technology partners Accenture, MHRA established its Sentinel IT system in 2005 which is used by most agency business areas to manage business processes for:

- marketing authorisations;
- pharmacovigilance;
- clinical trials;
- manufacturer's and wholesale dealer's licences;
- inspections;
- issuing GMP and GDP certificates and automatic loading of these into the EMA's EudraGMDP database.

In February 2013 a newly developed Sentinel risk information module was deployed to expand upon the paper-based RBI system initiated in 2009.

The Risk Estimation Tool uses the intelligence data collected on regulated companies, their respective sites and previous inspection results across all GxP areas to predict a risk score as "likely next inspection result". This score is calculated for every site and can be interpreted as a weighted sum of inspection findings. Companies/sites are ranked based on predicted risk and business rules are applied to suggest a next inspection date.

A planning step allows inspectors to accept or reject the suggested date taking into account other information which may not be included in the statistical calculation. For estimation of the risk score, the tool uses a logistic regression statistical model incorporating all data elements for all companies and sites. The model is fit (i.e. recomputed) monthly based on the most recent data extracted from Sentinel. The Empirica algorithm software was designed by Oracle Health Services to provide detailed analysis of the risk information. MHRA first used Empirica software in 2006 for pharmacovigilance signal detection and management.

The model estimates the association between inspection findings and other covariates (events) observed in data. The algorithm makes a global estimate on how these events affect inspection score within a GxP and then applies this when these events are recorded in the future. As a result those factors which are statistically most relevant to risk will receive the highest weighting and this will be continuously updated as more events are recorded. The model looks at events over a five-year period but applies greater significance to more recent data.

Current implementation status

A number of aspects of the algorithm are being validated including:

- the risk score;
- the weighting of inspection outcomes;
- the weighting of the risk events;
- the generation of proposed inspection dates from the risk score;
- inclusion of all appropriate risk events.

The algorithm is being assessed on an individual GxP basis as well as across the GxPs. The algorithm output is being compared against the existing RBI processes within the GxPs. Until the algorithm has been successfully validated the existing risk-based inspection scheduling processes will remain in place.

GDP risk-based inspection (RBI) programme

The GDP risk-based inspection process commenced for all wholesale dealer's licence holders on 1 April 2009.

Compliance report

Sites will be required to complete a Compliance Report in advance of inspection: this will be prompted by the inspector. Guidance to completing the report can be found within the document. The Compliance Report should be returned to your inspector prior to the inspection.

Risk rating process

Inspectors use the inspection outputs along with a number of other factors to identify a risk rating for the site, which equates to a future inspection frequency. As this process is not concluded until the inspection is closed the risk ratings **will not** be discussed at the closing meetings. However a copy of the full inspection report which includes the full risk rating rationale is provided to sites once the inspection has been closed.

Issue of a certificate of GDP compliance and/or support of the site on the relevant licence is indication of meeting the minimum level of GDP compliance. Risk ratings identify the degree of surveillance required within the licensing and inspection program. There is no intention that sites be rated against each other as a result of risk ratings assigned by MHRA. Risk ratings can change following inspection resulting in either increased or decreased risk. Inspection risk ratings will not be published by MHRA.

There will be no formal process of appeal against risk ratings and future inspection frequency. However any rating that results in an increased inspection frequency from the previous standard will be peer reviewed before conclusion by a GDP operations manager. MHRA does have a formal complaints process if sites wish to log an issue; however, any concerns regarding the inspection process should be raised with the inspector.

Conditions of Holding a Wholesale Dealer's Licence

The holder of a wholesale dealer's licence must comply with certain conditions in relation to the wholesale distribution of medicinal products. These conditions are set out in regulations 43–45 of the Human Medicines Regulations 2012 [SI 2012/1916] ("the Regulations"). They require that the licence holder shall:

- comply with the guidelines on Good Distribution Practice;[2]
- ensure, within the limits of their responsibility as a distributor of medicinal products, the appropriate and continued supply of such medicinal products to pharmacies and persons who may lawfully sell such products by retail or who may lawfully supply them in circumstances corresponding to retail sale, so that the needs of patients in the UK are met;
- provide and maintain such staff, premises, equipment and facilities for the handling, storage and distribution of the medicinal products under the licence as are necessary to maintain the quality of, and ensure proper distribution of the medicinal products;
- inform the licensing authority of any proposed structural alteration to, or discontinued use of, premises to which the licence relates or premises which have been approved by the licensing authority;
- inform the licensing authority of any change to the Responsible Person;
- not sell or offer for sale or supply any medicinal product unless there is a marketing authorisation, Article 126a authorisation, certificate of registration or traditional herbal registration ("an authorisation") for the time being in force in respect of that product; and the sale or offer for sale is in accordance with the provisions of that authorisation. This restriction on the holder of a wholesale dealer's licence shall not apply to:

[2] Guidelines on Good Distribution Practice of Medicinal Products for Human Use (2013/C 343/01) http://eur-lex.europa.eu/LexUriServ/LexUriServ.do?uri=OJ:C:2013:343:0001:0014:EN:PDF.

- the sale or offer for sale of a special medicinal product; and
- the export to an EEA State, or supply for the purposes of such export, of a medicinal product which may be placed on the market in that State without a marketing authorisation, Article 126a authorisation, certificate of registration or traditional herbal registration by virtue of legislation adopted by that State under Article 5(1) of the 2001 Directive;
- the sale or supply, or offer for sale or supply, of an unauthorised medicinal product where the Secretary of State has temporarily authorised the distribution of the product under regulation 174 of the Regulations; or
- the wholesale distribution of medicinal products to a person in a third country.

The holder of a wholesale dealer's licence shall:

- keep such documents relating to the sale of medicinal products to which their licence relates as will facilitate the withdrawal or recall from sale of medicinal products in accordance with an emergency plan referred to below;
- have in place an emergency plan which will ensure effective implementation of the recall from the market of any relevant medicinal products where such recall is:
  - ordered by the licensing authority or by the competent authority of any other EEA State, or
  - carried out in co-operation with the manufacturer of, or the holder of the marketing authorisation for, the product in question;
- keep records in relation to the receipt, dispatch or brokering of medicinal products, of the date of receipt, the date of despatch, the date of brokering, the name of the medicinal product, the quantity of the product received, dispatched or brokered, the name and address of the person from whom the products were received or to whom they are dispatched, and the batch number of medicinal products bearing safety features referred to in point (o) of Article 54 of the 2001 Directive.[3]

Where the holder of a wholesale dealer's licence imports from another EEA State for which they are not the holder of the marketing authorisation, Article 126a authorisation, certificate of registration or a traditional herbal registration of the product, then they shall notify the holder of that authorisation of their intention to import that product. In the case where the product is the subject of a marketing authorisation granted under Regulation (EC) No 726/2004, the holder of the wholesale dealer's licence shall notify the EMA or for any other authorisation they shall notify the

[3] Point (o) of Article 54 was inserted by Directive 2011/62/EU of the European Parliament and of the Council (OJ No L 174, 1.7.2011, p.74).

licensing authority. In both cases they will be required to pay a fee to the EMA in accordance with Article 76(4) of the 2001 Directive[4] or the licensing authority as the case may be, in accordance with the Fees Regulations. These requirements will not apply in relation to the wholesale distribution of medicinal products to a person in a non-EEA country.

The licence holder, for the purposes of enabling the licensing authority to determine whether there are grounds for suspending, revoking or varying the licence, must permit a person authorised in writing by the licensing authority, on production of identification, to carry out any inspection, or to take any samples or copies, which an inspector could carry out or take under Part 16 (enforcement) of the Regulations.

The holder of a wholesale dealer's licence must verify that any medicinal products they receive which are required by Article 54a of the Directive[5] to bear safety features are not falsified. This does not apply in relation to the distribution of medicinal products received from a third country by a person for supply to a person in a third country. Any verification is carried out by checking the safety features on the outer packaging, in accordance with the requirements laid down in the delegated acts adopted under Article 54a(2) of the 2001 Directive.

The licence holder must maintain a quality system setting out responsibilities, processes and risk management measures in relation to their activities.

The licence holder must also immediately inform the licensing authority and, where applicable, the marketing authorisation holder, of medicinal products which the licence holder receives or is offered which the licence holder knows or suspects, or has reasonable grounds for knowing or suspecting, to be falsified.

Where the medicinal product is obtained through brokering, the licence holder must verify that the broker involved fulfils the requirements set out in the Regulations.

The licence holder must not obtain supplies of medicinal products from anyone except the holder of a manufacturer's licence or wholesale dealer's licence in relation to products of that description or the person who holds an authorisation granted by another EEA State authorising the manufacture of products of the description or their distribution by way of wholesale dealing. The supply must be in accordance with the principles and guidelines of good distribution practice. This does not apply in relation to the distribution of medicinal products directly received from a non-EEA country but not imported into the EU.

[4] Article 76(4) was inserted by Directive 2011/62/EU of the European Parliament and of the Council (OJ No L 174, 1.7.2011, p.74).

[5] Article 54a was inserted by Directive 2011/62/EU of the European Parliament and of the Council (OJ No L 174, 1.7.2011, p.74).

From 28 October 2013, where the medicinal product is directly received from a non-EEA country for export to a non-EEA country, the licensed wholesale dealer must check that the supplier of the medicinal product in the exporting non-EEA country is authorised or entitled to supply such medicinal products in accordance with the legal and administrative provisions in that country.

The holder of a wholesale dealer's licence must verify that the wholesale dealer who supplies the product complies with the principles and guidelines of good distribution practices, or the manufacturer or importer who supplies the product holds a manufacturing authorisation.

The holder of a wholesale dealer's licence may distribute medicinal products by way of wholesale dealing only to the holder of a wholesale dealer's licence relating to those products, the holder of an authorisation granted by the competent authority of another EEA State authorising the supply of those products by way of wholesale dealing, a person who may lawfully sell those products by retail or may lawfully supply them in circumstances corresponding to retail sale, or a person who may lawfully administer those products. This does not apply in relation to medicinal products which are distributed by way of wholesale dealing to a person in a non-EEA country.

From 28 October 2013, where the medicinal product is supplied directly to persons in a non-EEA country the licensed wholesale dealer must check that the person who receives it is authorised or entitled to receive medicinal products for wholesale distribution or supply to the public in accordance with the applicable legal and administrative provisions of the non-EEA country concerned.

Where any medicinal product is supplied to any person who may lawfully sell those products by retail or who may lawfully supply them in circumstances corresponding to retail sale, the licence holder shall enclose with the product a document which makes it possible to ascertain:

- the date on which the supply took place;
- the name and pharmaceutical form of the product supplied;
- the quantity of product supplied;
- the names and addresses of the person or persons from whom the products were supplied to the licence holder; and
- the batch number of the medicinal products bearing the safety features referred to in point (o) of Article 54 of the 2001 Directive.

The holder of a wholesale dealer's licence shall keep a record of the information supplied where any medicinal product is supplied to any person who may lawfully sell those products by retail or who may lawfully supply them in circumstances corresponding to retail sale for a minimum period of five years after the date on which it is supplied and ensure, during that period, that that record is available to the licensing authority for inspection.

The wholesale dealer's licence holder shall at all times have at their disposal the services of a responsible person who, in the opinion of the licensing authority, has knowledge of the activities to be carried out and of the procedures to be performed under the licence which is adequate for performing the functions of responsible person; and has experience in those procedures and activities which is adequate for those purposes.

The functions of the responsible person shall be to ensure, in relation to medicinal products, that the conditions under which the licence has been granted have been, and are being, complied with and the quality of medicinal products which are being handled by the wholesale dealer's licence holder are being maintained in accordance with the requirements of the marketing authorisations, Article 126a authorisations, certificates of registration or traditional herbal registrations applicable to those products.

The standard provisions for wholesale dealer's licences, that is, those provisions which may be included in all licences unless the licence specifically provides otherwise, insofar as those licences relate to relevant medicinal products, shall be those provisions set out in Part 4 of Schedule 4 of the Regulations.

The licence holder shall not use any premises for the purpose of the handling, storage or distribution of relevant medicinal products other than those specified in their licence or notified to the licensing authority by them and approved by the licensing authority.

The licence holder shall provide such information as may be requested by the licensing authority concerning the type and quantity of any relevant medicinal products which they handle, store or distribute.

Where and insofar as the licence relates to special medicinal products to which regulation 167 of the Regulations apply which do not have a UK or EMA authorisation and are commonly known as "specials" (refer to MHRA Guidance Note 14), the licence holder shall only import such products from another EEA State in response to an order which satisfies the requirements of regulation 167 of the Regulations; and where the following conditions are complied with:

- No later than 28 days prior to each importation of a special medicinal product, the licence holder shall give written notice to the licensing authority stating their intention to import that special medicinal product and stating the following particulars:
  - the name of the medicinal product, being the brand name or the common name, or the scientific name, and any name, if different, under which the medicinal product is to be sold or supplied in the United Kingdom;
  - any trademark or name of the manufacturer of the medicinal product,
  - in respect of each active constituent of the medicinal product, any international non-proprietary name or the British approved name or the monograph name or, where that constituent does not have an

international non-proprietary name, a British approved name or a monograph name, the accepted scientific name or any other name descriptive of the true nature of that constituent,
- the quantity of medicinal product which is to be imported which shall not exceed more, on any one occasion, than such amount as is sufficient for 25 single administrations, or for 25 courses of treatment where the amount imported is sufficient for a maximum of three months' treatment, and
- the name and address of the manufacturer or assembler of that medicinal product in the form in which it is to be imported and, if the person who will supply that medicinal product for importation is not the manufacturer or assembler, the name and address of such supplier.

- Subject to the next bullet point below, the licence holder shall not import the special medicinal product if, before the end of 28 days from the date on which the licensing authority sends or gives the licence holder an acknowledgement in writing by the licensing authority that they have received the notice referred to in the bullet point above, the licensing authority have notified them in writing that the product should not be imported.
- The licence holder may import the special medicinal product referred to in the notice where they have been notified in writing by the licensing authority, before the end of the 28-day period referred to in the bullet point above, that the special medicinal product may be imported.
- Where the licence holder sells or supplies special medicinal products, they shall, in addition to any other records which they are required to make by the provisions of their licence, make and maintain written records relating to the batch number of the batch of the product from which the sale or supply was made and details of any adverse reaction to the product so sold or supplied of which they become aware.
- The licence holder shall import no more on any one occasion than such amount as is sufficient for 25 single administrations, or for 25 courses of treatment where the amount imported is sufficient for a maximum of three months' treatment, and on any such occasion shall not import more than the quantity notified to the licensing authority in the notification of intention to import.
- The licence holder shall inform the licensing authority forthwith of any matter coming to their attention which might reasonably cause the licensing authority to believe that the medicinal product can no longer be regarded either as a product which can safely be administered to human beings or as a product which is of satisfactory quality for such administration.
- The licence holder shall not issue any advertisement, catalogue or circular relating to the special medicinal product or make any representations in respect of that product.

- The licence holder shall cease importing or supplying a special medicinal product if they have received a notice in writing from the licensing authority directing that, as from a date specified in that notice, a particular product or class of products shall no longer be imported or supplied.

The licence holder shall take all reasonable precautions and exercise all due diligence to ensure that any information they provide to the licensing authority which is relevant to an evaluation of the safety, quality or efficacy of any medicinal product for human use which they handle, store or distribute is not false or misleading in a material particular.

Where a wholesale dealer's licence relates to exempt advanced therapy medicinal products the licence holder shall keep the data for the system for the traceability of the advanced therapy medicinal products for such period, being a period of longer than 30 years, as may be specified by the licensing authority.

The Standard Provisions also require the holder of a wholesale dealer's licence that relates to exempt advanced therapy medicinal products to obtain supplies of exempt advanced therapy medicinal products only from the holder of a manufacturer's licence in respect of those products or the holder of a wholesale dealer's licence in respect of those products.

The licence holder must:

- distribute an exempt advanced therapy medicinal product by way of wholesale dealing only to the holder of a wholesale dealer's licence in respect of those products; or a person who may lawfully administer those products, and solicited the product for an individual patient;
- establish and maintain a system ensuring that the exempt advanced therapy medicinal product and its starting and raw materials, including all substances coming into contact with the cells or tissues it may contain, can be traced through the sourcing, manufacturing, packaging, storage, transport and delivery to the establishment where the product is used;
- inform the licensing authority of any adverse reaction to any exempt advanced therapy medicinal product supplied by the holder of the wholesale dealer's licence of which the holder is aware;
- keep the data for ensuring traceability for a minimum of 30 years after the expiry date of the exempt advanced therapy medicinal product or longer as specified by the licensing authority;
- ensure that the data for ensuring traceability will, in the event that the licence is suspended, revoked or withdrawn or the licence holder becomes bankrupt or insolvent, be held available to the licensing authority by the holder of a wholesale dealer's licence for the same period that the data has to be kept; and
- not import or export any exempt advanced therapy medicinal product.

Appointment and Duties of the Responsible Person

Title VII of the Directive on the Community code relating to medicinal products for human use (Directive 2001/83/EC) obliges holders of a distribution authorisation to have a "qualified person designated as responsible". Regulation 45 of the Human Medicines Regulations 2012 [SI 2012/1916] states the requirement for a Responsible Person (RP) within the UK.

The RP is responsible for safeguarding product users against potential hazards arising from poor distribution practices as a result, for example, of purchasing suspect products, poor storage or failure to establish the bona fides of purchasers. The duties of an RP include:

- to ensure that the provisions of the licence are observed;
- to ensure that the guidelines on Good Distribution Practice (GDP) are complied with;
- to ensure that the operations do not compromise the quality of medicines;
- to ensure that an adequate quality system is established and maintained;
- to oversee audit of the quality system and to carry out independent audits;
- to ensure that adequate records are maintained;
- to ensure that all personnel are trained;
- to ensure full and prompt cooperation with marketing authorisation holders in the event of recalls.

In order to carry out his or her duties, the RP should be resident in the UK and have a clear reporting line to the licence holder or MD. The RP should have personal knowledge of the products traded under the licence and the conditions necessary for their safe storage and distribution. The RP should have access to all areas, sites, stores and records which relate to the licensable activities and regularly review and monitor all such areas, etc. and the standards achieved.

If the RP is not adequately carrying out those duties, the licensing authority may consider the suspension of the licence, withdrawal of acceptance of the RP on that licence and the acceptability on any other licence.

The RP does not have to be an employee of the licence holder but must be available to the licence holder when required. Where the RP is not an employee, there should be a written contract specifying responsibilities, duties, authority and so on.

In the case of small companies, the licensing authority may accept the licence holder as the nominated RP. In larger companies, however, this is not desirable.

There is no statutory requirement for the RP to be a pharmacist.

The RP should have access to pharmaceutical knowledge and advice when it is required, and have personal knowledge of:

- the relevant provisions of the Human Medicines Regulations 2012 [SI 2012/1916];
- Directive 2001/83/EC as amended on the wholesale distribution of medicinal products for human use;
- the EU Guidelines on Good Distribution Practice of Medicinal Products for Human Use (2013/C 343/01);
- the conditions of the wholesale dealer's licence for which nominated;
- the products traded under the licence and the conditions necessary for their safe storage and distribution;
- the categories of persons to whom products may be distributed.

Where the RP is not a pharmacist or eligible to act as a Qualified Person (QP) (as defined in Directive 2001/83/EC as amended), the RP should have at least one year's practical experience in both or either of the following areas:

- Handling, storage and distribution of medicinal products.
- Transactions in or selling or procuring medicinal products. In addition, the RP should have at least one year's managerial experience in controlling and directing the wholesale distribution of medicinal products on a scale, and of a kind, appropriate to the licence for which nominated.

To carry out responsibilities, the RP should:

- have a clear reporting line to either the licence holder or the Managing Director;
- have access to all areas, sites, stores, staff and records relating to the licensable activities being carried out;
- demonstrate regular review and monitoring of all such areas, sites and staff etc. or have delegated arrangements whereby the RP receives written reports that such delegated actions have been carried out on behalf of the RP in compliance with standard operating procedures and GDP. Where arrangements are delegated, the RP remains responsible and should personally carry out the delegated functions at least once a year;
- focus on the management of licensable activities, the accuracy and quality of records, compliance with standard operating procedures and GDP, the quality of handling and storage equipment and facilities, and the standards achieved;
- keep appropriate records relating to the discharge of the RP responsibilities.

Where the licence covers a number of sites, the RP may have a nominated deputy with appropriate reporting and delegating arrangements. However, the RP should be able to demonstrate to the licensing authority that the necessary controls and checks are in place.

The licence holder should ensure that there is a written standard operating procedure for receiving advice and comment from the RP and recording the consequent action taken as may be necessary.

Should it prove impossible to resolve a disagreement between the licence holder and the RP, the licensing authority should be consulted.

While a joint referral is clearly to be preferred, either party may approach the licensing authority independently. If an RP finds difficulty over performing statutory responsibilities or the activities being carried out under the licence, the licensing authority should be consulted in strict confidence.

The Responsible Person Gold Standard

The Human Medicines Regulations 2012 require holders of a wholesale dealer's licence to designate and ensure that there is available at all times at least one person, referred to in the regulations as the "responsible person", who in the opinion of the licensing authority:

(**a**) has knowledge of the activities to be carried out and of the procedures to be performed under the licence; and
(**b**) has adequate experience relating to those activities and procedures.

Guidance on the role and responsibilities of the Responsible Person (RP) is set out in Chapter 2 of the EU GDP guidelines and these remain the same irrespective of whether the RP is a permanent or contracted employee of a company.

The RP plays a vital part in ensuring the quality and the integrity of medicinal products are maintained throughout the distribution chain and it is essential that they have the right knowledge, demonstrate competence and deploy the right skills so that patients and healthcare professionals have the confidence and trust to use medicines.

In order to facilitate this and to standardise the requirements for individuals operating as, or aspiring to be, a RP, Cogent (the national skills body for the science industries) has, following extensive discussion with pharmaceutical companies and MHRA, published a new Gold Standard role profile for the RP.

This sets out an industry-agreed framework that identifies the skills required in four competency areas and includes not only traditional qualifications and technical requirements but also the behavioural skills necessary to do the job to a high standard.

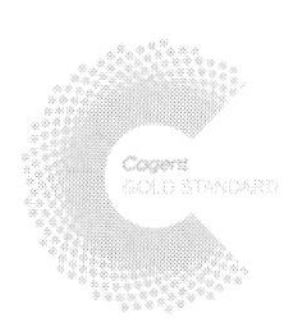

Responsible Person Medicinal Products

The Human Medicines Regulations require a distributor to designate a Responsible Person(s), named on the applicable licence. Regulation 45 and the EU GDP Guide set out the requirements and the responsibilities.	
Where the RP is contracted to a company, the duties remain the same as for those of the permanently employed RP. The responsibilities should be covered in a contract.	
Compliance	**The Gold Standard** Job Role skills, knowledge and behaviours
	the individual should understand: • *the role of MHRA in the licensing of medicines and as the competent authority including the risk-based inspection process, the role of the enforcement group, the Inspection Action Group (IAG), and resulting actions that can be taken due to non-compliance* • *the UK regulations in relation to Wholesale Distribution* • *the European Pharmaceutical Directive related to Wholesale Distribution of Medicinal Products* • *Good Distribution Practice (GDP)* • *the importance of a clear reporting line to the wholesale distribution authorisation holder, senior manager and/or CEO* *the individual shall:* • *employ due diligence in the discharge of their duties, maintaining full compliance to procedures and appropriate regulations* • *report to senior management, the Marketing Authorisation holder and the MHRA any suspicious events of which they become aware* *in addition, the individual also has knowledge of:* • *the role of the professional bodies and organisations that regulate those supplying medicinal products to the public e.g. GPhC* • *the role of the Home Office in relation to the handling of Controlled Drugs* • *the role of the Veterinary Medicines Directorate (VMD) in relation to veterinary medicines* • *the role of the European Medicines Agency (EMA) and use of EUDRAGMDP* • *the Falsified Medicines Directive* • *the Principles and Guidelines of Good Manufacturing Practice and how the principles of GDP maintain product quality throughout the distribution chain*
Knowledge	**The Gold Standard** Job Role skills, knowledge and behaviours
	the individual should have: • *the prior relevant knowledge and experience related to the distribution of medicinal products* • *access to pharmaceutical knowledge and advice when it is required* • *knowledge of the products traded under the licence* • *if not a pharmacist or QP, one year's relevant practical and managerial experience of medicinal products*

Responsible Person Medicinal Products

Technical Competence	The Gold Standard Job Role skills, knowledge and behaviours
	the individual is able to perform duties including: ***Quality Management*** *the individual shall ensure that a quality management system proportionate to the distributor's activities is implemented and maintained including:* • *Quality Risk Management* • *Corrective and Preventative Actions (CAPA) to address deviations* • *Change Control* • *Measurement of performance indicators and management review* ***Personnel*** *The Responsible Person is required to:* • *understand their own responsibilities* • *carry out all duties in such a way as to ensure that the wholesale distributor can demonstrate GDP compliance* • *define personal and staff roles, responsibilities and accountabilities and record all delegated duties* • *ensure that initial and continuous training programmes are implemented and maintained* • *ensure all personnel are trained in GDP, their own duties, product identification, the risks of falsified medicines and specific training for products requiring more stringent handling* • *maintain training records for self and others and ensure training is periodically assessed* ***Premises & Equipment*** • *ensure that appropriate standards of GDP are maintained for own premises and contracted storage premises* • *identify medicinal products, legal categories, storage conditions and different Marketing Authorisation types* • *maintain the safety and security of medicinal products within the appropriate environments, including product integrity and product storage* • *use the appropriate systems to segregate, store and distribute medicinal products* • *maintain records for the repair, maintenance, calibration and validation of equipment including computerised systems* • *ensure storage areas are temperature mapped, qualified and validated* ***Documentation*** *The individual shall focus on:* • *the accuracy and quality of records* • *contemporaneous records* • *records storage* • *maintaining comprehensive written procedures that are understood and followed* • *ensure procedures are valid and version controlled*

Supported by MHRA

Responsible Person Medicinal Products

	Operations • *carry out due diligence checks and ensure that suppliers and customers are qualified* • *ensure all necessary checks are carried out and that medicinal products are authorised for sale* • *manage authorised activities to ensure operations do not compromise the quality of medicines and can demonstrate compliance with GDP* • *demonstrate the application of activities and provisions in accordance with the wholesale distribution authorisation and of company processes and procedures* • *ensure that any additional requirements imposed on certain products by national law are adhered to e.g. specials, unlicensed imports & Controlled Drugs* ***Complaints, returns, suspected falsified medicinal products and medicinal product recalls*** • *ensure relevant customer complaints are dealt with effectively, informing the manufacturer and/or marketing authorisation holder of any product quality/product defect issues* • *decide on the final disposition of returned, rejected, recalled or falsified products* • *approve any returns to saleable stock* • *coordinate and promptly perform any recall operations for medicinal products* • *co-operate with marketing authorisation holders and national competent authorities in the event of recalls* • *have an awareness of the issues surrounding falsified medicines* ***Outsourced Activities*** • *approve any subcontracted activities which may impact on GDP* ***Self-Inspection*** • *ensure that self-inspections are performed at appropriate regular intervals following a prearranged programme and necessary corrective measures are put in place* ***Transportation*** • *apply the appropriate transport requirements and methods for cold chain, ambient and hazardous product* • *ensure all transport equipment is appropriately qualified* ***Brokers*** • *ensure that transactions are only made with brokers who are registered* • *ensure that any broker activities performed are registered*
Business Improvement	**The Gold Standard** Job Role skills, knowledge and behaviours
	the individual should: • *practise continuous improvement practices and utilise appropriate tools and techniques to solve problems*

Supported by MHRA

Responsible Person Medicinal Products

Functional & Behavioural	**The Gold Standard** Job Role skills, knowledge and behaviours

	the individual has:
	• *relevant skills in:*
	  ○ *English (level 2)*
	  ○ *Mathematics (level 2)*
	  ○ *ICT*
	
	the individual can demonstrate relevant personal qualities in:
	Autonomy
	  ○ *take responsibility for planning and developing courses of action, including responsibility for the work of others*
	  ○ *exercise autonomy and judgement within broad but generally well-defined parameters*
	• *Management & Leadership*
	  ○ *develop and implement operational plans for their area of responsibility*
	  ○ *manage diversity & discrimination issues*
	  ○ *provide leadership for their team*
	• *Working with others*
	  ○ *ensure effective delegation whilst retaining ownership of the outcome*
	  ○ *develop and maintain productive working relationships with colleagues and stakeholders*
	  ○ *monitor the progress and quality of work within their area of responsibility*
	• *Personal development*
	  ○ *manage their professional development by setting targets and planning how they will be met*
	  ○ *review progress towards targets and establish evidence of achievements*
	• *Communication*
	  ○ *put across ideas in clear and concise manner and present a well-structured case*
	  ○ *communicate complex information to others*
	• *Business*
	  ○ *understands the business environment in which the company operates*
	  ○ *has an appreciation of the industry sector and competitors*
	• *Customers*
	  ○ *understands the customer base and is aware of customer requirements*

For more information on how to achieve the Gold Standard contact us on 0845 607 014

Version 2 July 2014

Supported by MHRA 

Quality Management

Introduction

A consistent focus on quality is of prime importance for all wholesale distributors in order to maintain an effective and efficient business that meets customer needs and ensures product quality is maintained.

Quality management is as much a mind-set as an activity in itself, with the aim of achieving quality processes that permeate throughout all distribution activities, with the ultimate goal of ensuring patient safety.

Quality system

The quality system is the vehicle by which quality management is delivered and it should be all encompassing. However, this doesn't mean that it has to be complicated with its size and complexity being proportionate to the distribution activities being undertaken.

Chapter 1 of the EU GDP Guidelines gives a detailed breakdown of the areas that should be covered.

Quality Risk Management (QRM)

QRM is not new to wholesale distributors, as informal systems have always been in place as a matter of routine. However, GDP now require that the process be documented.

QRM is essentially the identification and control of risks to product quality through the evaluation of the activities that are being performed.

Underpinning this is that the evaluation of the risk is based on knowledge and experience of the process and ultimately links to the protection of the patient.

The level of effort, formality and documentation of the process should be commensurate with the level of risk.

GDP refers to ICH Q9 as a useful guidance document on QRM. This includes the principles and concepts of QRM with Annex 1 & 2 identifying various tools and techniques: www.ich.org

Change control

Change control is a formal process whereby changes to a process are identified, planned and introduced in a controlled way.

A typical change control process may consist of several stages:

- Submission of a documented change control request to management.

- Consideration and evaluation of the request by management. This evaluation should consider if the change poses any risks to public health or if there are any impacts on other processes and procedures. Actions identified at this stage should be resolved prior to approval.
- Implementation of the change. This is a live process and unforeseen events during implementation should be subject to a further assessment of risk.
- Review of effectiveness. After an appropriate period an objective review of the effectiveness of the change should take place. This should consider both successes as well as any problems that have arisen, with any lessons learnt being used to improve the process for future changes.

Deviation management and Corrective and Preventative Actions (CAPA)

Distributors should prepare for circumstances when things go wrong.

When this occurs the identification of control measures should be a priority, particularly where an unplanned event may have (or has had) a significant impact on public health.

Depending on the severity of the event, there may be a process of ongoing control, requiring initial corrective actions to deal with the immediate fallout from the event through to the identification of the root cause of the problem and proactive preventative actions to avoid reoccurrence.

During investigations it is very important that the root cause of the problem is identified, as what may initially be deemed as for example, "operator error" may actually be attributable to system weaknesses elsewhere.

Corrective and Preventative Actions (CAPA), are an integral part of the change control and deviation management processes, reducing the likelihood or severity of a risk event and creating an opportunity to improve and reduce the likelihood of the deviation occurring again.

Corrective actions are identified after a deviation with preventative actions being identified following an event and before a change is implemented.

Management review and monitoring

To ensure the effectiveness and continual improvement of the quality management system, there should be a process in place for regular management review.

This review should actively involve company management and key personnel such as the Responsible Person.

In very small organisations the process of review may require taking a step back and objectively reviewing the operational activities and supporting procedures.

Questions to be considered during the management review may include:

- Is the quality system effective, does it reflect the current business model?
- Are activities being carried out in accordance with the principles of Quality Management, or are there opportunities for improvement with some processes requiring modification?
- Has the regulatory landscape changed? Has legislation been amended or guidance published which means that activities need to be reviewed?
- Is the business operating effectively, are the right products getting to the right customers at the right time? What improvements can be made to address customer complaints?
- If third parties are contracted to provide services, are they complying with their contractual obligations?
- Have there been any unforeseen problems or events and has effective CAPA been implemented?
- Have planned changes been effective?
- Are there staff training or staff re-training issues?

The outcome of each management review should be documented, result in a CAPA plan where necessary and should be effectively communicated to staff.

Controls on Certain Medicinal Products

The EU Guidelines on Good Distribution Practice (GDP) require the licence holder and/or the RP to consider extra requirements throughout the wholesale process. The RP must ensure that any additional requirements imposed on certain products by national law are adhered to. This will include ensuring that the correct licences are held for the activities undertaken, ensuring any broker used or any organisation performing outsourced activities are correctly licensed and that customers are entitled to receive the products supplied. Where products are exported there may be additional national requirements to consider. Supplies made to customers should be monitored to enable any irregularity in the sales patterns of narcotics, psychotropic substances and other dangerous substances to be investigated. Unusual sales patterns may constitute diversion and must be reported to MHRA.

Some products may present additional risks during storage and transport. These could include cytotoxics, radiopharmaceuticals and flammable products. Special destruction requirements may be imposed by national regulation.

Staff may require additional training for the receipt and handling of hazardous products, radioactive materials, products presenting special risks of abuse as well as temperature-sensitive products. Handling of temperature-sensitive products is addressed elsewhere in this publication.

Transaction arrangements, storage, transport and security measures should all be documented, risk assessed and audited to demonstrate continued suitability and compliance.

GDP frequently refers to 'additional requirements imposed on certain products by national law'. These medicinal products and substances are subject to additional controls not frequently encountered by the licensed wholesale dealer. Article 83 of 2001/83/EC permits member states to have more stringent requirements for:

- narcotic or psychotropic substances;
- medicinal products derived from blood;
- immunological medicinal products;
- radiopharmaceuticals.

These additional controls are managed by MHRA or other organisations. Some examples are given below.

Controlled drugs and precursor chemicals

The Home Office Drug Licensing and Compliance Unit issues licences for those who handle controlled drugs and precursor chemicals. These licences cover handling of all scheduled medicines, some unscheduled medicines and some Active Substances. It is your responsibility to make sure that you have the appropriate licence for the activities you are undertaking with controlled substances. Home Office requirements may change over time, for example, in 2013 further controls were applied to exports of pseudoephedrine and ephedrine and in 2014 several medicines were reclassified.

Licences are specific to a legal person (including corporate entities), and if your company operates at a number of sites you will need to ensure you have the appropriate licence for each site.

Guidance and information on obtaining the correct licence are provided on the GOV.UK website.[6,7]

[6] https://www.gov.uk/government/uploads/system/uploads/attachment_data/file/118341/fees-guidance.pdf.

[7] https://www.gov.uk/government/uploads/system/uploads/attachment_data/file/275027/precursor_regulations_changes_dec_2013.pdf.

Use of the mandatory requisition form for Schedule 2 and 3 controlled drugs

On 30 November 2015 legislative provisions came into effect which made it mandatory for specified health and veterinary care professionals, and organisations listed at Regulation 14(4) of the Misuse of Drugs Regulations 2001, to use an approved form for the requisitioning of Schedule 2 and 3 controlled drugs. This change, which the Home Office consulted on in 2011, implemented a final recommendation of the Shipman Inquiry on requisitions.[8]

Following the introduction of the form, the Home Office has been made aware that activities within the hospital sector, which would normally be governed by provisions under Regulation 14(6), and which were not expected to come within the scope of the new requirement, is now captured as a result of changes in the NHS structures in recent years. This is an unintended consequence of the changes to NHS structures and healthcare delivery since 2011 rather than a result of a regulatory change.

The Home Office circular which introduced the change made it clear that the requirement to use the mandatory form was to be limited to activities undertaken by health and veterinary care practitioners in the community to enable their requisition activities to be monitored. The requirement to use a mandatory form was not expected to extend to the hospital environment, where traditionally supplies of controlled drugs were undertaken by an onsite pharmacy owned by the hospital and under the Regulation 14(6) provisions.

This additional guidance is therefore being issued to further explain how the provisions governing the use of the new mandatory form may be interpreted. This guidance does not impact on the need for a Home Office licence. The guidance should also not be interpreted as a change in Home Office licensing requirements, which these regulatory and procedural requirements relate to but are not necessarily interdependent on.

(1) The use of the form is mandatory when individual health and veterinary care professionals requisition the relevant controlled drugs in the community, including when such drugs are ordered from pharmaceutical wholesalers and community pharmacies.
(2) It is not the intention of the policy on requisitions forms to capture requisition activities for hospital wards, etc. However, as a number of hospitals wards now obtain the relevant controlled drugs from other trusts or organisations, and therefore across legal entities, these activities do fall within the scope of Regulation 14(2) and therefore

[8] https://www.gov.uk/government/publications/circular-0272015-approved-mandatory-requisition-form-and-home-office-approved-wording/circular-0272015-approved-mandatory-requisition-form-and-home-office-approved-wording.

the use of the form is mandatory, unless an exemption applies (e.g. registered pharmacies are completely excluded from the requirement to use a requisition when obtaining the relevant controlled drugs). In order to ensure that the regulatory provisions are complied with, it is the view of the Home Office that the person in charge or acting person in charge of a hospital or care home (excluding hospices) can issue a "bulk" or "global" requisition based on previous year's orders to a separate legal entity which supplies its wards. Hospital or Trust wards can then draw on this "bulk" requisition throughout the year using the duplicate order forms as happens presently. Hospital or Trust wards therefore do not need to use the mandatory form when obtaining the relevant controlled drugs from another trust or legal entity. Similarly, the person in charge or acting person in charge of an Ambulance Trust, as defined under the 2001 Regulations, can issue a "bulk" requisition when obtaining the relevant controlled drugs to be supplied directly to employees of the Trust by a separate legal entity. However, those employees, when individually drawing on the "bulk" requisition, must use the mandatory form.

(3) Additionally, pharmaceutical wholesale suppliers (excluding community pharmacies) are exempt under current regulatory provisions from submitting requisitions received for Schedule 2 and 3 controlled drugs to the NHS Business Services Authority (NHSBSA). The only requisition forms that must be submitted to the NHSBSA following the introduction of the mandatory form are those provided by individual healthcare professionals when obtaining the relevant controlled drugs in the community.

The table below provides some examples of how the regulations apply in specific circumstances with reference to Regulations 14(2) and (4). This is only a Home Office view and does not constitute legal advice. Organisations are advised to seek their own independent legal advice where appropriate. This guidance has been developed with the Department of Health and the Care Quality Commission.

Supplier	Recipient	Do I need to use the new FP10CDF?	Do I need to submit the FP10 CDF to the NHSBSA
Wholesaler	Practitioners Paramedics and the other professionals and organisations listed in Regulation 14(4)	Yes, all practitioners and the list of healthcare professionals must use the form when they requisition for stocks	No, the Regulation exempts the submission of requisitions received by wholesalers from being sent to the NHSBSA
Wholesaler	Registered Pharmacy (including a hospital registered pharmacy)	NO	N/A
Wholesaler	Hospital, care home or ambulance trust without a registered pharmacy **Excludes Hospices	Yes, in line with Regulation 14	No, the regulation exempts the submission of requisitions received by wholesalers from being sent to the NHSBSA
Legal entity A	Legal entity B's hospital wards	Yes, person in charge of legal entity B must issue a yearly global requisition to legal entity A. Wards in legal entity B can then draw on this requisition through the year using duplicate order books	NO
Legal entity A	Legal entity B's registered pharmacy	NO	N/A
Legal entity A	Legal entity B	Yes if organisation is listed at Regulation 14(4)	NO

Controls on strategic goods and drugs useable in execution by lethal injection

The Export Control Organisation (ECO), part of the Department for Business, Innovation and Skills, issues licences for controlling the export of strategic goods. The range of goods and services covered by embargoes and sanctions is extensive, changes from time to time and includes:

- military equipment;
- dual-use goods that can be used for both civil and military purposes;
- products used for torture;
- radioactive sources.

These may apply to medical devices, finished medical products or Active Substances.

ECO also administers the European Union controls on trade in certain goods which could be used for capital punishment, torture or other cruel, inhuman or degrading treatment or punishment. This includes the control on the export of certain drugs usable in execution by lethal injection.

As a result of these EU-imposed controls, exporters need to seek appropriate permission from national export control authorities to export to any destination outside the EU, short and intermediate acting barbiturate anaesthetic agents including, but not limited to, the following:

- amobarbital (CAS RN 57-43-2);
- amobarbital sodium salt (CAS RN 64-43-7);
- pentobarbital (CAS RN 76-74-4);
- pentobarbital sodium salt (CAS 57-33-0);
- secobarbital (CAS RN 76-73-3);
- secobarbital sodium salt (CAS RN 309-43-3);
- thiopental (CAS RN 76-75-5);
- thiopental sodium salt (CAS RN 71-73-8), also known as thiopentone sodium.

The control also applies to products containing one or more of the above. These controls are intended to apply to finished products – in other words, those that are packaged for human or veterinary use. It is not intended that they should apply to raw materials or to intermediate products (i.e. products that require further processing to make them suitable for human or veterinary use).

It is the responsibility of a wholesale dealer to ensure that they export responsibly and within the law. ECO provide training for exporters of strategic goods. Further information on training and licensing can be found on the GOV.UK website.[9,10]

Control of lisdexamfetamine, tramadol, zaleplon, zopiclone and reclassification of ketamine

On 10 June 2014, the Parliamentary Order controlling and, in the case of ketamine, reclassifying the following drugs, came into force:

- lisdexamfetamine;
- tramadol;
- zaleplon;

[9] https://www.gov.uk/controls-on-torture-goods.

[10] https://www.gov.uk/government/organisations/export-control-organisation.

- zopiclone;
- medicines containing these substances.

The Order, available on legislation.gov.uk, controls:

- lisdexamfetamine as a Class B drug;
- tramadol as a Class C drug;
- zopiclone and zaleplon as Class C drugs.

The Order also reclassifies ketamine as a Class B drug under the Misuse of Drugs Act 1971.

Companies who possess, supply or produce lisdexamfetamine, tramadol, zaleplon, zopiclone (or medicines containing these substances) need to get the correct licences from the Home Office.

More information on how to apply for a licence, and how much they cost are available on www.GOV.UK at https://www.gov.uk/controlled-drugs-licences-fees-and-returns or by calling the Duty Compliance Officer on 020 7035 8972.

Companies without the correct licences are at risk of prosecution.

The listed drugs will be scheduled, alongside their control, as follows to ensure that they remain available for use in healthcare:

- Lisdexamfetamine (a drug which converts to dexamfetamine when administered orally and used as second-line treatment for ADHD) will be listed in Schedule 2 alongside dexamfetamine.
- Tramadol will be listed in Schedule 3 but exempted from the safe custody requirements. Full prescription writing requirements under Regulation 15 will apply to its use in healthcare.
- Zopiclone and zaleplon will be listed in Part 1 of Schedule 4 alongside zolpidem.

Ketamine is not being rescheduled immediately. In line with the Advisory Council on the Misuse of Drugs' (ACMD) advice, the Home Office will carry out a public consultation later this year to assess the impact of rescheduling ketamine to Schedule 2.

A final decision on the appropriate schedule for ketamine will be made after the consultation. Until then ketamine will remain a Schedule 4 Part 1 drug.

Home Office Circular 008/2014: A change to the Misuse of Drugs Act 1971 – Control of NBOMes, Benzofurans, Lisdexamphetamine, Tramadol, Zopiclone, Zaleplon and Reclassification of Ketamine.[11]

[11] https://www.gov.uk/government/publications/circular-0082014-changes-to-the-misuse-of-drugs-act-1971.

Controls and authorisations applying to those handling medicinal products derived from blood

MHRA is responsible for the controls and authorisations that apply to blood establishments (BE) and controls that apply to hospital blood banks (HBB) and sites that collect, test and supply human blood or blood components intended for transfusion. Further information can be obtained from the GOV.UK website.[12]

Procurement, storage, distribution and disposal of vaccines

EU GDP applies to all vaccine wholesale activities. The supply chain in the UK and guidance on who may receive these, how the end user should store them and how they should be disposed of are further described in *Immunisation against infectious disease*, commonly known as the green book. This can be found on the GOV.UK website.[13]

Wholesale of radiopharmaceuticals

Many activities will fall under EU GMP Annex 3 as well as EU GDP. Further guidance from the marketing authorisation must be followed.

Wholesale of veterinary medicinal products

The Veterinary Medicines Regulations 2013 [SI 2013/2033] came into force on 1 October 2013. A wholesaler of veterinary products will usually be required to hold a WDA(V). The regulations require that the holder must:

(a) store veterinary medicinal products in accordance with the terms of the marketing authorisation for each product;
(b) comply with the Guidelines on Good Distribution Practice of Medicinal Products for Human Use as if the veterinary medicinal products were authorised human medicinal products;
(c) carry out a detailed stock audit at least once a year; and
(d) supply information and samples to the Secretary of State on demand.

[12] https://www.gov.uk/guidance/blood-authorisations-and-safety-reporting.

[13] https://www.gov.uk/government/collections/immunisation-against-infectious-disease-the-green-book#the-green-book.

Further information in relation to import and export, the cascade, record keeping, controlled drugs and other veterinary wholesale activities can be found on the GOV.UK website.[14]

Sourcing and Exporting Medicinal Products – Non-EEA Countries

The Falsified Medicine Directive[15] describes the concept that medicinal products may be introduced into the Union which are not intended to be released for free circulation.[16] These are known as introduced medicinal products or products sourced from a non-EEA country for the specific reason of re-exporting them back to a non-EEA country.

The Directive goes on to report that:

"*The provisions applicable to the export of medicinal products from the Union and those applicable to the introduction of medicinal products into the Union with the sole purpose of exporting them need to be clarified. Under Directive 2001/83/EC a person exporting medicinal products is a wholesale distributor. The provisions applicable to wholesale distributors as well as good distribution practices should apply to all those activities whenever they are performed on Union territory, including in areas such as free trade zones or free warehouses.*"[17]

Consequently the Falsified Medicine Directive extends Title VII of the Directive 2001/83/EC on the need for a wholesale dealer's licence[18] to medicine exported from the UK to non-EEA countries and to the sourcing and re-export of medicine from and to non-EEA countries by way of wholesale with the introduction of Article 85a.

Article 85a of the amended Directive 2001/83/EC provides the clarification needed by removing certain wholesale dealing obligations, leaving the remainder of the obligations to apply. It also adds further specific obligations to conduct certain checks on both supplier and customer in the non-EEA countries and makes it clear that the existing provisions for providing paperwork apply.

> *Article 85a*
>
> In the case of wholesale distribution of medicinal products to third countries, Article 76 and point (c) of the first paragraph of Article 80 shall not apply. Moreover, points (b) and (ca) of the first paragraph

[14] https://www.gov.uk/government/organisations/veterinary-medicines-directorate.

[15] Directive 2011/62/EU on preventing the entry of falsified medicinal products into the legal supply chain.

[16] Recital 10.

[17] Recital 15.

[18] An Article 77(1) authorisation.

of Article 80 shall not apply where a product is directly received from a third country but not imported. However, in that case wholesale distributors shall ensure that the medicinal products are obtained only from persons who are authorised or entitled to supply medicinal products in accordance with the applicable legal and administrative provisions of the third country concerned. Where wholesale distributors supply medicinal products to persons in third countries, they shall ensure that such supplies are only made to persons who are authorised or entitled to receive medicinal products for wholesale distribution or supply to the public in accordance with the applicable legal and administrative provisions of the third country concerned. The requirements set out in Article 82 shall apply to the supply of medicinal products to persons in third countries authorised or entitled to supply medicinal products to the public.

Introduced medicinal products

For a medicinal product to be an introduced medicinal product it has to be sourced from a non-EEA country by a licensed wholesale dealer and re-exported back to a non-EEA country by the same licensed wholesale dealer. An introduced medicinal product will not have a marketing authorisation for the UK or another EEA country. Such products can be held on a licensed site in a free zone or in a bonded warehouse under an appropriate Customs processing procedure.

A medicinal product that is sourced from a non-EEA country by a licensed wholesale dealer, for the purpose of supply to another legal entity in the UK or the EEA, is not an introduced medicinal product as it will have been freely circulated to that other legal entity within the community when supplied. In such circumstances the activity will be subject to the need for a manufacturer's licence authorising import as required under Article 40 of Directive 2001/83/EC. A medicinal product, which is not introduced, will also be subject to the requirements set out in "Title III – Placing on the market" of Directive 2001/83/EC, unless an exemption exists.

When dealing with introduced medicinal products a licensed wholesale dealer has to ensure that they only source the medicine from a person in the non-EEA country who is authorised or entitled to supply medicinal products for wholesale distribution in accordance with the applicable legal and administrative provisions of the non-EEA country concerned. A licensed wholesale dealer should also verify that the medicinal products received are not falsified.

A licensed wholesale dealer also has to ensure that they only supply the introduced medicinal product back to a person in the non-EEA country who is authorised or entitled to receive medicinal products for wholesale distribution or supply to the public in accordance with the applicable legal and administrative provisions of the non-EEA country concerned.

A separate Home Office licence is required for introduced medicines that are also "controlled drugs".

The introduction of a medicinal product can be subdivided into the following categories.

PHYSICAL INTRODUCTION

Introduced medicines are sourced from a non-EEA country for the sole purpose of export to a non-EEA country. These products by definition will not have a marketing authorisation within the EEA. Under the provisions of a wholesale dealer's licence the medicinal product may not be "imported" that is the customs procedure code quoted on the C88 cannot include sole or simultaneous entry into free circulation within the EEA. A suitable customs procedure code should be declared.

Generally the site of *holding* these products prior to export will be a registered customs warehouse that is the subject of a wholesale dealer's licence.

Further information in respect of the Union Customs Code (UCC) introduced across the European Union on 1 May 2016 may be obtained from HMRC.

FINANCIAL INTRODUCTION

Any trade between two non-EEA countries being facilitated and invoiced from a Member State will be subject to the EU Good Distribution Practice (GDP) Guidelines in their entirety.

It is necessary for the supplying licensed wholesale dealer to demonstrate full compliance with EU GDP Guidelines regardless of the fact that the medicinal product does not physically enter an EEA country.

Companies operating this business model will be required to have a wholesale distribution authorisation authorising the activities of *Procurement* and *Supply* as they are buying and selling the product.

This activity should not be confused with brokering. A broker does not at any time take ownership of the product; a broker will bring a buyer and seller together and typically will receive a commission or fee from one or both parties. Brokers are subject to a registration requirement. They must have a permanent address and contact details in the Member State in which they are registered. Brokers are discussed in more detail elsewhere in this guide. See Chapters 6–9.

Export to a non-EEA country

A medicinal product that is exported to a non-EEA country will be supplied to the specification of the importing country concerned. This might be a medicinal product that is:

- licensed in the UK and has a national marketing authorisation issued by MHRA;
- licensed in another EEA member state and has a national marketing authorisation issued by the competent authority for medicines of the EEA member state concerned;
- a medicine that is the subject of a centralised marketing authorisation issued by the European Medicines Agency;
- an unlicensed medicine manufactured specifically for export by the holder of a manufacturer's licence;[19]
- an introduced product that the same authorised wholesale dealer has sourced from a non-EEA country;
- a medicine that is also a "controlled drug" – requiring a separate Home Office licence. (See section titled 'Controls on certain medicinal products' earlier in this chapter.)

An unlicensed medicine know as a "special" manufactured by the holder of a manufacturer's specials licence, cannot be distributed to a non-EEA country. This is because to accord with Directive 2001/83/EC an unlicensed medicine manufactured for export has to be manufactured by the holder of an Article 40 of Directive 2001/83/EC manufacturing authorisation and certified prior to release by the Qualified Person named on that manufacturing authorisation.

A licensed wholesale dealer that exports a medicine that is the subject of an EEA national or central marketing authorisation to a non-EEA country must obtain their supplies of medicinal products only from persons who are themselves in possession of an EEA wholesale distribution authorisation[20] or an EEA manufacturer's authorisation.[21]

They must also verify that the authorised medicinal product they receive is not falsified by checking the safety feature on the outer packaging in accordance with the requirements laid down in the Commission Delegated Regulation (EU) 2016/161.[22]

[19] Article 40 of Directive 2001/83/EC authorisation.

[20] Article 77 of Directive 2001/83/EC authorisation.

[21] Article 40 of Directive 2001/83/EC authorisation.

[22] Supplementing Directive 2001/83/EC of the European Parliament and of the Council by laying down detailed rules for the safety features appearing on the packaging of medicinal products for human use.

A licensed wholesale dealer also has to ensure that they supply the authorised medicine only to persons in the non-EEA country who are authorised or entitled to receive medicinal products for wholesale distribution or supply to the public in accordance with the applicable legal and administrative provisions of the non-EEA country concerned.

The definition of wholesale distribution does not depend on whether the distributor is established or operating in specific customs areas, such as in free zones or in free warehouses. A wholesale distribution authorisation is required and all obligations related to wholesale distribution activities also apply to these distributors.

A number of companies and their sites that were not previously regulated now require a wholesale distribution authorisation. Parties that may be affected include freight consolidators, freight forwarders and logistics services providers in the air, sea or road transport sector when they are "holding" medicinal products.

Customs-approved warehouse facilities "holding" medicinal products are also required to have a wholesale distribution authorisation.

To clear a medicinal product for export, HMRC will need to know exactly what is being shipped. Medicinal products presented for export should therefore be accompanied by the **commercial invoice** fully detailing the commodity to be exported. Use of "proforma invoices" for export purposes are to be avoided.

Customs documentation must be completed fully and legibly. A false or misleading declaration may lead to a fine or to seizure of the item.

The products may be subject to restrictions in the country of destination or there may be trade embargoes or other restrictions preventing certain medicinal products from being exported to certain destinations.

It is the responsibility of the exporter to enquire into import and export regulations (prohibitions, restrictions such as quarantine, pharmaceutical restrictions, etc.) and to find out what documents, if any (commercial invoice, certificate of origin, health certificate, export licence, authorisation for goods subject to quarantine (plant, animal, food products, etc.)) are required in the destination country. See EU GDP 2013/C 343/01. Chapter 5, Section 5.9:

> "Where wholesale distributors supply medicinal products to persons in third countries, they shall ensure that such supplies are only made to persons who are authorised or entitled to receive medicinal products for wholesale distribution or supply to the public in accordance with the applicable legal and administrative provisions of the country concerned."

Export is defined as to *allow Community goods to leave the customs territory of the Union*. It is the responsibility of the exporter (consignor) to ensure that:

- medicinal products remain in the legal supply chain;
- export paperwork is completed correctly;
- proof of export is obtained and retained;
- the Consignee is entitled to receive the medicinal products;
- the product has been stored, handled and shipped in accordance with GDP.

HMRC recommends that exporters should routinely provide their freight agents with the following:

- Their UK Economic Operator Registration and Identification (EORI) number.
- Details of whom the goods are to be consigned to, their name and address in full.
- A commercial reference that can be incorporated into the Declaration Unique Consignment Reference (DUCR) to assist with the export audit trail.
- Details of where the goods are to be exported, i.e. country of final destination.
- Shipping or flight details (where known).
- Correct value of goods in correct currency code.
- The Commodity Code and a clear and unambiguous description of the goods, their quantity marks and numbers.
- Any reference numbers already issued by HMRC, for example, Inward Processing Relief, Outward Processing relief authorisations or previous declarations should also be provided.

Qualification check in non-EEA countries

MHRA recognises that conducting a qualification check in some non-EEA countries can be quite difficult. In the first instance suppliers and customers in the non-EEA countries concerned should be asked to justify their local entitlement and requirements. However an authorised wholesale dealer should not just accept information from the supplier or customer that they can supply or import medicines, but should take reasonable steps to verify that the information that they have been provided with is valid and accurate. This may involve making additional checks with the regulatory authority for medicines in the country concerned where possible and documenting the outcome as evidence. Any licences obtained should have been translated into English and authenticated by a notary with appropriate due diligence carried out. MHRA's expectation is that an authorised wholesale dealer should have oversight of the export of the introduced medicinal product and have all the appropriate documentation to evidence it. See also section on Qualification of Customers and Suppliers.

Wholesale obligations for exported and introduced medicinal products

For further obligations of licensed wholesale dealers, see Chapters 2 and 3 on Legislation on Wholesale Distribution.

Incoterms®

Incoterms® 2010 rules are internationally accepted standard definitions of trade terms. Incoterms® were developed by the International Chamber of Commerce, Paris, France in 1936 and have been regularly revised to reflect the changes in transportation and documentation. The current version is Incoterms® 2010.

Full information may be obtained from www.iccwbo.org.

There are 11 Incoterms® which can be broadly split into subgroups according to which party pays for and arranges the main transport. Each group is discussed briefly below.

The first three groups are applicable to products moved by Air, Road, Rail or Sea, or by a combination of transport modes.

BUYER ARRANGES MAIN CARRIAGE – EXW

- EXW – EX Works (... named place).

This Incoterm® puts the onus on the customer to complete and pay for the export formalities and for the pre-carriage of the products from the seller's premises to the point of export.

In many cross-border and international transactions this presents practical difficulties. Specifically, the exporter needs to be involved in export reporting formalities and cannot realistically leave these to the buyer.

BUYER ARRANGES MAIN CARRIAGE – FCA

- FCA – Free Carrier (... named place of delivery).

With this Incoterm® the seller has control regarding the transport and customs formalities prior to the goods leaving the Union; the buyer pays for the main transport.

This option crucially gives the seller control over the pre transport to the point of export and control over the export formalities.

It is the legal responsibility of the declarant to ensure that the goods are accurately declared and presented to HMRC prior to goods leaving the UK. If the exporter is employing a freight agent to declare the goods on his behalf he must ensure he supplies them with the appropriate information to submit a legal declaration.

SELLER ARRANGES MAIN CARRIAGE: CPT; CIP; DAT; DAP; DDP

- CPT – Carriage Paid To (... named place of destination);
- CIP – Carriage and Insurance Paid To (... named place of destination);
- DAT – Delivered At Terminal (... named terminal at destination port);
- DAP – Delivered At Place (... named place of destination);
- DDP – Delivered Duty Paid (... named place of destination).

Here the seller has control over the pre-carriage, customs formalities and the main transport mode.

The following two groups are only applicable when transport is via sea freight or by inland waterways.

BUYER ARRANGES MAIN CARRIAGE: FAS; FOB

- FAS – Free Alongside Ship (... named port of shipment);
- FOB – Free On Board (... named port of shipment).

Pre-carriage and export formalities by seller the customer pays for the sea freight from the named port of shipment.

SELLER ARRANGES MAIN CARRIAGE: CFR; CIF

- CFR – Cost and Freight (... named port of destination);
- CIF – Cost Insurance and Freight (... named port of destination).

Pre-carriage and export formalities by seller; seller also pays sea freight to the named destination.

When entering into trade negotiations with potential clients, the selection of the appropriate Incoterm® can significantly influence the ability to demonstrate GDP compliance.

Selection of an Incoterm® such as ExW or FCA where the buyer pays for the freight does not absolve the seller of his GDP responsibilities with respect to selection of transport conditions.

As the Consignor, the exporter is legally responsible for the information shown on the customs documentation. The definition of export is *to allow Community goods to leave the customs territory of the Union.*

Import of medicinal products

Import is not a function authorised under a wholesale distribution authorisation. For medicinal products which are imported from a non-EEA country, an Article 40 authorisation (Manufacturer's/Importer's Authorisation (MIA)) is required.

Import is briefly discussed in this section for the sake of completeness and to make a clear distinction between the act of Introduction, a GDP activity executed under a wholesale dealer's licence and Importation, a GMP activity requiring an MIA.

Importation of product manufactured in a third country for use in a Member State is a GMP activity. This requires a Manufacturing Licence (typically a Manufacturer's/Importer's Authorisation but, in particular circumstances, a Manufacturer's Specials licence may be required). Import activity associated with an MIA requires Q.P. release of product. Full details of these activities may be found in the publication *Rules and Guidance for Pharmaceutical Manufacturers and Distributors* (The Orange Guide).

Transactions where title is passed from EEA company to non-EEA company

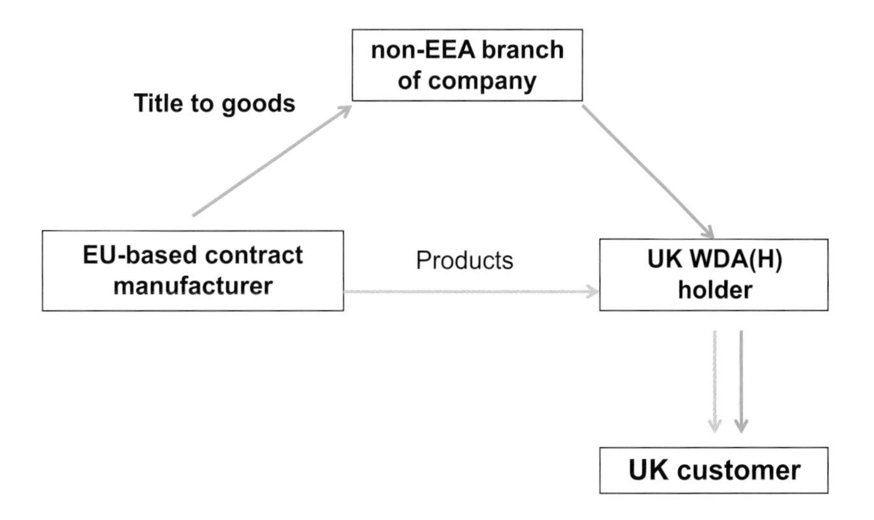

In this scenario, licensed medicinal product is manufactured and released within a Member State under a Manufacturing Authorisation. The title of the goods is transferred to the ***sole UK pre-distribution partner*** via a third country while the product physically remains within the EU at suitably licensed premises at all times.

Product transfers directly from the contract manufacturer to the ***sole UK pre-distribution partner***. The pre-distributor partner then distributes the products according to their own authorisation.

To accord with Regulation 44(1) of the Human Medicines Regulations the holder of a wholesale dealer's licence may not obtain supplies of medicinal products from anyone except:

(**a**) the holder of a manufacturer's licence or wholesale dealer's licence in relation to products of that description; or
(**b**) a person who holds an authorisation granted by another EEA State authorising the manufacture of products of that description or their distribution by way of wholesale dealing.

Therefore these requirements will prohibit the purchasing of medicines from a company in a third country. This transaction is classed as a ***fiscal import*** from a third country and therefore is not WDA(H) activity. This business model requires an MIA.

Purchasing from a Member State is not classed as import and is simply trade within the Union.

Ensuring you have the correct authorisations for the intended activity

Directive 2001/83/EC defines wholesale distribution of medicinal products to mean:

> "All activities consisting of procuring, holding, supplying or exporting medicinal products, other than supplying these products to the public."[23]

An authorised wholesale dealer may be authorised to undertake one or more of these activities according to the scope of the granted authorisation.

Definitions of these activities are provided in the Annex to EU Guidelines on Good Distribution Practice of Medicinal Products for Human Use (GDP)[24] as below:

> **PROCURING** – obtaining, acquiring, purchasing or buying medicinal products from manufactures, importers or other wholesale distributors.
> **HOLDING** – storing medicinal products. (*MHRA apply the interpretation that a site is deemed to be holding when either an*

[23] Article 1(17).
[24] 2013/C 343/01.

ambient product is on site for more than 36 hours or there is active refrigeration taking place regardless of dwell time.)

SUPPLYING – all activities of providing, selling, donating medicinal products to wholesalers, pharmacists or persons authorised or entitled to supply medicinal products to the public.

EXPORT– to allow Community goods to leave the customs territory of the Union. For the purposes of Directive 2013/C 343/01 the supply of medicines from EU Member State to a contracting State of the European Economic Area is not considered as export.

Trade with a company based in another Member State does not constitute export. An authorised wholesale dealer may trade with an authorised wholesale dealer in another Member State or contracting State of the EEA once appropriate initial and ongoing verification of the customer and/or supplier status has been established.

An authorised wholesale dealer may be authorised to trade in medicinal products in one or more of these categories:

- With a Marketing Authorisation in EEA country(s).
- Without a Marketing Authorisation in the EEA and intended for EEA market.
- Without a Marketing Authorisation in the EEA and intended for Exportation.

This third category defines introduced medicinal products.

Appendix 4 sets out the type of licence required for the sourcing of a medicinal product, distribution with in the UK and EEA, and for the export from the UK, including introduced medicine.

A Home Office licence is required to import or export "controlled drugs". Controlled drugs are named in the Misuse of Drugs legislation, and grouped in schedules. Anyone intending to supply a controlled drug must apply for the relevant schedule licence.[25]

Temperature Control and Monitoring

Manufacturers subject their products to stability studies that are used to determine appropriate storage conditions including those for temperature. These conditions are therefore specific for each product, and a licensed wholesale dealer should refer to manufacturers' information when deciding the storage conditions to use.

[25] https://www.gov.uk/guidance/controlled-drugs-licences-fees-and-returns.

Following manufacture, some medicinal products can be stored and transported at ambient temperature, whilst others may require lower than ambient temperatures to assure their quality and efficacy.

These are often referred to as "cold chain products" or "fridge lines" and wholesale dealers are expected to store and distribute them in strict accordance with the product labelling requirements as stated in the EU GDP Guidelines – Chapters 5.5 (Storage) and 9.2 (Transportation) give more information.

Medicinal products experiencing an adverse temperature may undergo physical, chemical or microbiological degradation. In the most serious of cases this may lead to conversion of the medicine to ineffective or harmful forms. The ability to detect these changes may not appear until the medicine is consumed, and it is therefore essential that appropriate temperature conditions are controlled and monitored throughout each step of the supply chain. This section concerns temperature mapping and the ongoing temperature monitoring and control required throughout the wholesale supply chain.

Temperature mapping

Chapter 3.2.1 of the GDP Guidelines states:

> "An initial temperature mapping exercise should be carried out on the storage area before use, under representative conditions.
> Temperature monitoring equipment should be located according to the results of the mapping exercise, ensuring that monitoring devices are positioned in the areas that experience the extremes of fluctuations.
> The mapping exercise should be repeated according to the results of a risk assessment exercise or whenever significant modifications are made to the facility or the temperature controlling equipment.
> For small premises of a few square meters which are at room temperature, an assessment of potential risks (e.g. heaters) should be conducted and temperature monitors placed accordingly."

Whilst the guidelines say that mapping should take place, it does not state how this should be done.

Temperature mapping should be carried out to demonstrate by way of documented evidence that the chosen storage area is suitable for the storage of temperature-sensitive medicinal products. A mapping exercise of the proposed storage area will also ensure that the company understands their storage area and has identified any potential areas therein that may be unsuitable to store medicines. A mapping exercise will also inform as to where permanent thermometers should be located.

Temperature mapping should be carried out, if possible, before stock is stored. This might not be possible where a storage area is being reconfigured. In smaller empty storage areas, dummy products could be used to simulate normal operational storage without compromising genuine product, including cold stores and fridges/freezers. In an empty storage area, a mapping exercise should be repeated when fully stocked. Data arising from the exercise should be documented and a risk assessment documented with any hot or cold spots identified. This exercise should then be repeated to take into account seasonal variations.

To temperature map, firstly look at the area to be used for storage and identify the highest point of storage not the highest shelf or pallet location. Identify any potential problem areas such as heaters, lighting, windows and doors, loading bays or high storage areas such as mezzanine floors. These areas should be covered in the exercise. Areas such as CD rooms, packing areas, returns and quarantine should be included. When deciding on a storage area, it can be difficult to cool storage areas down as well as heat them up. Calibrated monitoring probes should be used in sufficient numbers dependent on the size of the storage area.

Once the initial mapping exercise is complete, the data should be recorded and risk assessed to determine the most appropriate positions for the permanent monitoring probes and should cover the areas that have the widest temperature fluctuations or indicate areas with any hot or cold spots. A risk assessment would also define and justify the regularity of any future mapping exercises and must also be regularly reviewed, perhaps as part of the self-audit process.

The exercise should be repeated to cover seasonal variations or if the storage area is subsequently reconfigured.

The RP should be party to the whole mapping process and should be fully aware of the mapping exercise findings, risk assessment recommendations and review process. The RP's involvement does not stop at the mapping process, however; the RP should also be able to demonstrate supervision and review of subsequent daily minimum/maximum routine temperature monitoring and recording and should be consulted in the event of any temperature excursions.

Refrigerated and ambient medicinal products, receipt, storage, packing and transportation

RECEIPT OF REFRIGERATED PRODUCTS

When cold chain products are received, it is important that they are checked-in as a matter of priority and placed in a pharmaceutical refrigerator.

The person responsible for receiving the delivery must also satisfy themselves that the goods have been transported under appropriate conditions (e.g. there has been no direct contact between the products and gel or ice blocks or if the consignment is warm to the touch).

If it cannot be confirmed that the products have been transported under appropriate conditions and there is concern that their quality may have been compromised, the delivery should be quarantined in a suitable refrigerator while enquiries are made with the supplier.

Until the issue has been clarified the products in question should be considered as unsuitable and should not be supplied.

If, following enquiries, there is still doubt as to the quality of the medicines received, the delivery should not be accepted and should be returned to the supplier.

STORAGE OF REFRIGERATED PRODUCTS IN A PHARMACEUTICAL REFRIGERATOR

A pharmaceutical refrigerator is required for the storage of refrigerated medicinal products.

The air within this type of refrigerator is circulated by a fan, which provides a uniform temperature profile and a rapid temperature pull down after the door has been opened.

Temperature monitoring is recorded by a calibrated electronic min/max thermometer, with an accuracy of ±0.5°C, which can be read without opening the refrigerator door.

Additional benefits are that these refrigerators can be locked and some have the option of either an audio or visual alarm system to alert staff in the event of temperature deviations.

Many refrigerators have glass fronted doors giving greater visibility of stock levels, aiding stock management and also deterring the storage of non-medicinal products.

When purchasing a new refrigerator, factors to consider might also include how long the unit can maintain the required temperatures if the power is turned off and to what extent the temperature is affected by external ambient temperature variation, for example, in hot spells.

TEMPERATURE MONITORING IN A REFRIGERATOR

As is applicable for transportation, products stored in a refrigerator should be subject to daily temperature monitoring by a minimum and maximum calibrated device with a supporting appropriate calibration certificate.

Temperature records should identify any temperature deviations and give details of corrective actions taken as a result.

For instances where there has been a temperature deviation, best practice would be to take a further reading later the same day, to ensure that it was a transient deviation and show that the temperature was now back within prescribed parameters.

The Responsible Person should be informed of any deviations.

Temperature records are especially important in the event of a problem with a product and may be required as evidence of appropriate storage. With this in mind, they should be free from alterations or corrections and the person responsible for taking the readings each day should have a trained deputy to cover for absences.

The records should be routinely reviewed and signed off by the Responsible Person.

SMALL REFRIGERATORS

Refrigerators used to store pharmaceuticals should be demonstrated to be fit for purpose. In the simplest of cases a new off-the-shelf refrigerator installed according to the manufacturer's instructions and temperature monitored with an appropriate device may be considered appropriately qualified for storing cold chain product that is shown to be unaffected by minor temperature excursions. A refrigerator used for holding more susceptible stock such as biological products will require more extensive qualification.

In addition to temperature mapping and monitoring there should be safeguards to preserve appropriate storage conditions. Some small refrigerators are purported to be medical or pharmaceutical refrigerators but this on its own does not automatically render them suitable for wholesale use. The refrigerator should be capable of restoring the temperature quickly after the door has been opened and without danger of overshooting to extreme cold. This could be assisted by an internal fan and good shelf design which enables an efficient air flow. There should be no internal ice box and no internal temperature dials capable of being inadvertently knocked and adjusted.

Storage practices for using small refrigerators should include consideration of segregation of stock with different status, e.g. incoming, quarantine, returned and outgoing stock. Sufficient space should be maintained to permit adequate air circulation and product should not be stored in contact with the walls or on the floor of the refrigerator. If the refrigerator is filled to capacity the effect on temperature distribution should be investigated. Where non-refrigerated items are introduced to the refrigerator, such as non-conditioned gel packs, the impact of introducing these items should be assessed regarding the increase in temperature they cause.

LARGE COMMERCIAL REFRIGERATORS AND WALK-IN COLD ROOMS

Large commercial refrigerators and walk-in cold rooms should be of appropriate design, suitably sited and be constructed with appropriate materials. The design should ensure general principles of GDP can be maintained, such as segregation of stock. Condensate from chillers should not be collected inside the unit and there should be a capability to carry out routine maintenance and service activities as much as possible from outside the unit. The temperature should be monitored with an electronic temperature-recording device that measures load temperature in one or more locations depending on the size of the unit, and alarms should be fitted to indicate power outages and temperature excursions.

FREEZERS

The same general principles apply to freezers as apply to other cold chain storage units above. Walk-in freezers pose a significant operator health and safety risk, and the impact of ways of working should be reviewed with consideration of risk to causing temperature excursions.

Control and monitoring of storage areas

Where medicines that may be required in an emergency are stored then contingency measures should be put in place such as linking essential equipment in a large warehouse to a source of emergency power. These emergency measures should be routinely tested, such as the confirmation of restoration of stored data and settings when emergency power supply is activated and after normal power is resumed. For these products there should be a system in place to ensure that on-call personnel are notified in the event of power failure or temperature alarms being triggered including notification outside of normal working hours.

Best practice

Whatever type of refrigerator or cold store is used, once a mapping exercise has taken place, products should be stored in an orderly fashion on shelves – not directly on the floor of the unit – to ensure air circulation, consistent temperatures throughout and to facilitate cleaning.

Calibrated temperature monitoring probes should be sited in a central location within the refrigerator and, preferably, between the products.

Probes should not be placed in the door.

The refrigerator should be cleaned regularly (as part of a general cleaning rota) and serviced at least annually.

If the refrigerator is fitted with an audible or visual alarm, this should be routinely tested to confirm correct operation at specified appropriate temperatures.

The stock within the refrigerator should be subject to effective stock rotation based on first expiry, first out (FEFO).

It should not be assumed that the most recent deliveries will have a longer expiry period.

Refrigerators containing medicinal products must not be used for the storage of food and drink or anything that might contaminate the medicinal products.

Calibration of temperature monitoring devices including ambient

In order to have confidence in temperature readings, monitoring devices should be calibrated to demonstrate they have appropriate accuracy and precision. Temperate storage thermometers should be capable of reading ±1°C, and cold chain devices capable of reading ±0.5°C. Calibration should extend across the whole of the working range, so for a temperate storage range of 15°C to 25°C the calibration range may be 10°C to 30°C to allow the thermometer to be used in assessing temperature excursions or to be used in temperature mapping exercises. Results of the calibration exercise should be presented in a report or calibration certificate approved by the calibrator and demonstrated to be appropriate for use by the wholesaler. The certificate should include the following details:

- serial number of the calibrated instrument;
- serial numbers of test instruments;
- traceability to national or international calibration standards;
- calibration test method used;
- ISO or equivalent registration details of calibration laboratory;
- date of calibration;
- calibration results;
- unique certificate number;
- approval of results by calibrator.

Where a temperature monitoring device reads the temperature from a main monitoring unit plus a remote probe it should be clear from the calibration certificate which part of the device the calibration refers to. Calibration should be carried out annually and, where adjustments are made to the equipment as part of calibration, an assessment of accuracy and precision should be made before and after adjustment. On completion a suitable representative from the wholesaler should approve the calibration indicating its suitability for use.

Packing of consignments and temperature management during transportation

Before being transported, refrigerated products should be packed in such a way as to ensure that the required temperatures are maintained throughout the journey and the medicines are transported in accordance with their labelling requirements to prevent jeopardising their quality. Chapter 9.3 of the EU GDP Guidelines mentions this. Additionally, the same chapter states:

> "Containers should bear labels providing sufficient information on handling and storage requirements and precautions to ensure that the products are properly handled and secured at all times. The containers should enable identification of the contents of the containers and the source."

For small volumes of lower risk products, with short journey times of less than three hours, validated insulated containers can be used.

For extended journeys, gel or ice packs are added to the packaging to maintain appropriate temperatures throughout. The positioning of these packs within the consignment is extremely important and they must not be allowed into direct contact with the products being shipped.

Bespoke packaging with compartments for the gel or ice blocks is available, although securely encasing the blocks in some form of wrapping, such as bubble wrap, or installing some other form of a buffer can be equally effective.

Larger volumes of refrigerated products will generally be shipped in refrigerated transport. This is particularly important if transportation times may be protracted or liable to delay.

Whatever method of transport is used, it is important to show that the required temperatures can be maintained. Chapter 9 of the EU GDP Guidelines is relevant.

Best practice is the implementation of temperature monitoring as a matter of routine for all refrigerated deliveries, but especially within shipments of high risk products. The temperature should be strictly controlled and monitored with calibrated temperature probes, to provide temperature data for the entire journey.

This data should be retained by the wholesaler for the required five years.

Daily minimum and maximum temperature monitoring and recording should also be carried out at all storage locations. Any recording devices should be calibrated.

Licensed wholesale dealers should review the temperature records or data for each consignment and there should be procedures in place for implementing corrective action in the case of adverse events. They should also ensure that consignments of refrigerated goods are clearly labelled with the required storage and transport conditions to be maintained as stated above.

The application of Mean Kinetic Temperature (MKT) to temperature monitoring of wholesale products is only appropriate where an acceptable MKT value is provided by the MA holder for a specific product, and the recording of temperature can be confirmed to be consistent and complete from the moment of leaving the manufacturer's premises. In practice the application of MKT fails where a complete chain of temperature recording cannot be allocated to a specific consignment of a product. Attempts to apply MKT have been proposed by wholesalers as an alternative to having adequate temperature control within their warehouses as well as attempting to downgrade the impact of temperature excursions. The use of MKT in the wholesale environment without robust supporting information and methodology is therefore discouraged.

A risk-based approach should be in place when planning transportation and should include the method of transport and delivery routes. EU GDP Guidelines, Chapter 1.5 states "Quality risk management is a systematic process for the assessment, control, communication and review of risks to the quality of medicinal products. It can be applied both proactively and retrospectively". ICH Q9 also gives examples of quality risk management (QRM) processes.

Some refrigerated items, such as vaccines, biotech products such as insulin and products derived from blood, plasma or serum, may be classified as high risk because they are vulnerable to freezing as well as exposure to high temperatures.

Other products, for example, chloramphenicol eye drops, may be labelled as requiring storage between 2°C and 8°C but a short deviation from this temperature range may present less of a danger to patients.

Following dispatch from a manufacturing facility, the distribution chain for medicinal products can be complex, potentially involving a number of storage locations, wholesalers and modes of transport, before the delivery finally reaches the patient.

The transportation arrangements from one location to another should be regarded as an extension of the storage activities and distributors are expected to treat each journey as unique with the length and complexity, as well as any seasonal variations, being considered when choosing the packing method and mode of distribution.

THIRD PARTY COURIERS

When licensed wholesalers employ couriers, they must satisfy themselves that they can adhere to the EU GDP Guidelines and can provide the service for which they are engaged.

The selection of third party service providers is very important and roles and responsibilities must be defined by a written contract.

Chapter 7 of the EU GDP Guidelines (Outsourced Activities) provides advice in this area and Chapter 1.3 covers the management of outsourced activities.

Handling Returns

Returns of refrigerated medicinal products

Because of the inherent dangers of returning refrigerated products, many licensed wholesale dealers will not consider refrigerated returns for subsequent resale in any event. All such returns are immediately stored in a dedicated and marked area awaiting collection by a licensed disposal company.

In the event of a licensed wholesale dealer accepting a return of a refrigerated product, possibly because of its high monetary value, the product should be returned in accordance with MHRA guidance (below), in an appropriate method of transport, with supporting documentation, such as a returns form. The returns form would normally include the reason for the return, contain details of the product and how it has been stored and should be signed by an authorised and identifiable signatory.

A trained person at the wholesalers should examine the returned product to check for tampering and to confirm that the return has been made in accordance with MHRA guidance. If this examination cannot be undertaken immediately, the product should be stored in a dedicated and marked area in a refrigerator until the checks can be made.

Provided the checks are satisfactory and are documented, the product may then be returned to saleable stock. MHRA guidance on managing returned non-defective (ambient) and refrigerated medicinal products is provided below.

Returns of non-defective medicinal products

Any person acting as a wholesale distributor must hold a wholesale dealer's licence.

Article 80(g) of Directive 2001/83/EC provides that distributors of human medicines must comply with the principles of and guidelines for good distribution practice (GDP).

The Commission has revised its guidelines for GDP which are now contained in the Guidelines of 5 November 2013 on Good Distribution Practice of Medicinal Products for Human Use (2013/C 343/01). Chapter 6.3 of the GDP Guidelines refers to returned medicinal products, the key elements being that:

"Medicinal products that have left the premises of the distributor should only be returned to saleable stock if ...:

(i) the medicinal products are in their unopened and undamaged secondary packaging and are in good condition; have not expired and have not been recalled;
...
(ii) it has been demonstrated by the customer that the medicinal products have been transported, stored and handled in compliance with their specific storage requirements;
(iii) they have been examined and assessed by a sufficiently trained and competent person authorised to do so;
..."

MHRA re-affirm that a licensed site can only be interpreted as being under full GDP control at a licensed wholesale dealer. This applies to all categories of medicines. Medicinal products held in unlicensed storage and distribution sites are not considered to be within the licensed wholesale distribution network.

AMBIENT RETURNS FROM A LICENSED WHOLESALE DEALER'S SITE

MHRA will adopt a pragmatic approach to the return of non-defective medicinal products for those products returned from a customer operating from a licensed wholesale dealer authorisation (WDA) site.

In such circumstances, the return should be completed as expeditiously as possible and the most expedient and appropriate method of transportation must be used.

The Responsible Person or the authorised person receiving the return, must be able to demonstrate evidence of "full knowledge" of the storage of the returned products throughout the period it has been with the customer, including transportation.

AMBIENT RETURNS FROM AN UNLICENSED SITE

For those non-defective ambient medicinal products returned from an unlicensed site, the return should be completed within five days, including transport.

The Responsible Person or the authorised person must be able to demonstrate evidence of "full knowledge" of the storage while at the unlicensed site, including transportation.

REFRIGERATED RETURNS FROM A LICENSED WHOLESALE DEALER'S SITE

MHRA will adopt a pragmatic approach to the return of non-defective medicinal products for those products returned from a customer operating from a licensed WDA site.

In such circumstances, the return should be completed expeditiously and the most expedient and appropriate method of transportation must be used.

The Responsible Person or the authorised person receiving the return, must be able to demonstrate evidence of "full knowledge" of the storage of the returned products throughout the period it has been with the customer, including transportation.

REFRIGERATED RETURNS FROM AN UNLICENSED SITE

For those non-defective refrigerated medicinal products returned from an unlicensed site, the return should be completed within 24 hours, including transport.

The Responsible Person or the authorised person must be able to demonstrate evidence of "full knowledge" of the storage while at the unlicensed site, including transportation.

Short-Term Storage of Ambient and Refrigerated Medicinal Products – Requirements for a Wholesale Dealer's Licence

The EU GDP Guidelines define wholesale distribution as:

> "... all activities consisting of procuring, holding, supplying or exporting medicinal products ..."

The Glossary of Terms defines holding as "storing medicinal products". Medicinal products should therefore only be stored on premises that are covered by a wholesale dealer's licence. However, there are certain cases where medicinal products are held for short periods of time during transportation and prior to onward shipment, e.g. in the transportation vehicle at motorway service stations or in overnight freight depots. In such instances it has been determined that, as a matter of policy, a site does not have to be named on a licence where ambient products are stored for less than 36 hours.

Sites holding ambient products in excess of 36 hours must be licensed. This applies only where ownership of the products has not been transferred to the person carrying out the storage activities. Where ownership has been transferred, this is supply and as such the receiving site must be licensed.

It is also important to note that, where wholesaling activities other than storage are being carried out, the site should be named on the relevant licence. This includes the handling of returned goods and where decisions are made regarding suitability for resale, as well as the usual activities of picking against orders.

Sites where refrigerated products are held, even when this is for less than 36 hours, must be licensed. The exception will be where these products are transported and stored overnight in continuously refrigerated vehicles.

The provisions of Chapter 9.2 of the EU GDP guidelines must also be observed. As with any delivery, staff receiving goods should also be alert for the presence of falsified medicines.

Sales Representative Samples

Under the legislation on advertising medicines, companies may only provide free samples to persons qualified to prescribe the medicine. Samples may only be supplied in response to a signed and dated request from the prescriber and must be appropriately labelled and accompanied by a copy of the Summary of Product Characteristics (SPC). The company must have adequate procedures for control and accountability for all samples. See section 6.12 of MHRA's Blue Guide for details of the legal requirements.

MHRA is aware that in some cases sales representatives receive samples to fill prescriber requests and that the medicinal products are delivered to them by colleagues or by couriers. Either way, the storage and delivery arrangements for these medicinal products must be validated to ensure the medicinal product will be transported expeditiously under controlled Good Distribution Practice (GDP) conditions and in accordance with labelled storage requirements at all times. It is highly unlikely that samples requiring refrigeration will meet these requirements. With regards to storage it is not acceptable for samples to be stored in the representative's home (on unlicensed premises, which are not GDP compliant), lacking appropriate storage facilities, security and controls to maintain the quality of the medicines and provide an audit trail.

Likewise distribution of samples involving delivery in a representative's vehicle that has no provision for maintaining correct storage conditions is also unacceptable. Temperatures in a car boot in high summer could reach 50 degrees Celsius or go below 0 degrees Celsius in winter. The practice of providing sales representatives with samples of medicinal products which they retain for onward distribution is therefore unlikely to be acceptable due to the storage and transport difficulties outlined above. The only reason for which sales representatives may hold samples on a long-term basis is for the purpose of product identification. In this regard procedures must be in place to ensure accountability for any such stock and to ensure no packs are provided to healthcare professionals.

Qualification of Customers and Suppliers

Before commencing wholesale dealing activities with a supplier or customer (trading partners), licensed wholesale dealers must ensure that their proposed trading partners are entitled to trade with them. Checks must demonstrate that trading partners either hold the required manufacturing and wholesale dealer's licence where necessary or that they are entitled to receive medicines for the purpose of retail supply, to a person who may lawfully administer the products or for use in the course of their business.

Qualification of suppliers

Maintaining the integrity of the supply chain is one of the most important aspects of wholesale distribution. A robust fully documented system to ensure medicines are sourced appropriately must be in place and subjected to regular review. Licensed wholesale dealers must ensure their suppliers are appropriately licensed to supply medicines. The qualification of suppliers requires the following steps to be fully compliant:

- The licence of the supplier should be viewed, either a copy obtained from the company or the details can be viewed on MHRA's website that has registers of licensed wholesale dealers and manufacturers.[26] Whilst MHRA registers are updated regularly they must not be relied on as a sole means of qualifying suppliers' authority to supply.
- The EudraGMDP website has daily updates from MHRA and contains all current live licences. Currently the EudraGMDP website does not contain details of the legal categories of products that can be handled or third party sites used and must not be relied on as a sole means of qualifying suppliers. When searching for companies use an asterisk (*) on either side of a name or number to broaden the search. You must also select UK as the country on the site details section. One practical way to demonstrate qualification of suppliers could be a printed copy of the appropriate pages, signed and dated as evidence the checks were made, when and by whom.[27]
- For supplies from other EEA Member States the same checks should be made on EudraGMDP and via licences that have been translated. The translated licences should be authenticated as such by a notary.

[26] http://www.mhra.gov.uk/Howweregulate/Medicines/Licensingofmedicines/Manufacturersandwholesaledealerslicences/index.htm.

[27] http://eudragmdp.ema.europa.eu/inspections/logonGeneralPublic.do.

COMPLIANCE WITH GDP

Licensed wholesale dealers must verify that wholesale suppliers comply with the principles and guidelines of good distribution practices. To establish GDP compliance, the GDP certificate of the wholesaler should be viewed on the EudraGMDP website. Certificates when issued are valid for five years. The date of the certificate expiry should be recorded. MHRA adds conditioning statements to GDP certificates for new applicants and those companies where the inspection outcome indicated a more frequent inspection schedule is required, limiting certificate expiry to two years. If there is no GDP certificate available then other evidence of GDP compliance by the wholesale supplier should be obtained, such as a copy of the inspection close-out letter confirming GDP compliance.

ROUTINE RE-QUALIFICATION

Licensed wholesale dealers must be aware of issues that could affect their suppliers' continued authority to supply. The following should be carried out:

- Regular checks at least twice a month of MHRA's list of suspended licence holders.
- Regular checks on EudraGMDP website for issued GMP and GDP statements of non-compliance.
- At least annually, a documented full re-qualification of suppliers.

DUE DILIGENCE

When entering into a new contract with new suppliers, the licensed wholesale dealer should carry out "due diligence" checks in order to assess the suitability, competence and reliability of the other party. Questions that should be considered could include the following:

- Checking the financial status of the supplier, how long have they been trading, and do they have an acceptable credit history?
- Has an audit been performed of the supplier, or has anyone in your company visited them? If so what was their impression?
- Where is the stock coming from and is the product offered a new product for your company?
- Is the product being offered available in quantities or volumes that are unusually high or is the price being offered lower than the usual price?
- How transparent is the supply chain of this transaction?
- What will be the method of transportation?

Due diligence checks should be implemented and documented when dealing with a company or transaction that is outside of an established trading pattern.

PROCUREMENT FROM THIRD COUNTRIES

For companies that are involved in the sourcing of medicines from a third country (a country outside of the European Economic Area) for export to third countries then there are different requirements. Companies are obliged to document that checks are made to show that where the medicinal product is directly received from a third country ("A") for export to a third country ("B"), the supplier of the medicinal product in country A is a person who is authorised or entitled to supply such medicinal products in accordance with the legal and administrative provisions in country A. Any licences obtained should have been translated into English and authenticated by a notary with appropriate due diligence carried out.

Customer qualification

Licensed wholesale dealers have a key role in guaranteeing medicines are only supplied to authorised organisations and qualified prescribers. For distribution to a wholesale customer, the checks that must be made are similar to the qualification of suppliers. In relation to wholesale distribution of scheduled controlled medicines to other wholesalers, companies must check their customers hold both a wholesale dealer's licence and a Home Office controlled drugs licence of the appropriate schedule.

For the qualification of a person who may lawfully administer the products then the following registers must be checked prior to supplying:

- Pharmacists and registered pharmacies – GPhC website register.
- Doctors – GMC list of registered medical practitioners.
- Dentists – General Dental Council list of registered dental practitioners.
- Paramedics/podiatrists/chiropodists – Health and Care Professions Council (HCPC) website register with listing for POM and local anaesthetic use, if these are the product categories sold.
- Practice nurses – Nursing and Midwifery Council register.
- Hospitals – the CQC in England or equivalents in Wales, Scotland and Northern Ireland.

See Appendix 3 for further information.

SUPPLY TO THIRD COUNTRIES

Licensed wholesale dealers exporting medicinal products to persons in third countries must ensure that such supplies are only made to persons who are authorised or entitled to receive medicinal products for wholesale distribution or supply to the public in accordance with the applicable legal and administrative provisions of the country concerned. As an example some companies attempt to export products such as Botox to doctors and clinics in the US. The US FDA only permits healthcare providers to obtain and use FDA-approved medications purchased directly from the manufacturer or from wholesale distributors licensed in the US. In certain circumstances, the FDA may authorise limited importation of medications that are in short supply. Such medications are imported from approved international sources and distributed in the US through a controlled network, and would not be sold in direct-to-clinic solicitations. UK-licensed versions of FDA-approved drugs are not treated by FDA as equivalent and must not be sold in the US to doctors or clinics.

DUE DILIGENCE

Licensed wholesale dealers have an obligation in GDP to monitor their transactions and investigate any irregularity in the sales patterns of narcotics, psychotropic or other dangerous substances. Unusual sales patterns that may constitute diversion or misuse of medicinal product should be investigated and reported to competent authorities where necessary.

If licensed wholesale dealers have any concerns about their customers ordering of controlled or other drugs, in the first instance they should email GDP.Inspectorate@mhra.gsi.gov.uk. Details of the company and the name and quantities of products that have been ordered in the last six months should be included. This will be dealt with in confidence.

ROUTINE RE-QUALIFICATION

It is important that there is periodic rechecking of the authority to receive medicines for your customers and such checks must be documented.

Falsified Medicines

A "falsified medicinal product" means any medicinal product with a false representation of:

(a) its identity, including its packaging and labelling, its name or its composition (other than any unintentional quality defect) as regards any of its ingredients including excipients and the strength of those ingredients;
(b) its source, including its manufacturer, its country of manufacturing, its country of origin or its marketing authorisation holder; or
(c) its history, including the records and documents relating to the distribution channels used.

The supply of falsified medicines is a global phenomenon and one which MHRA takes very seriously. Falsified medicines represent a threat to the legitimate UK supply chain and to patient safety. They are fraudulent and may be deliberately misrepresented with respect to identity, composition and/or source. Falsification can apply to both innovator and generic products, prescription and self-medication, as well as to traditional herbal remedies. Falsified medicines may include products with the correct ingredients but fake packaging, with the wrong ingredients, without active ingredients or with insufficient active ingredients, and may even contain harmful or poisonous substances.

The supply and distribution of medicines are tightly controlled within the European Community.

All licensed wholesalers must comply with the Community's agreed standards of good distribution practice (GDP) and there exist strict licensing and regulatory requirements in UK domestic legislation to safeguard patients against potential hazards arising from poor distribution practices: for example, purchasing suspect or falsified products, failing to establish the "bona fides" of suppliers and purchasers, inadequate record keeping, and so on.

Section 6.4 of the EU Guide to GDP is of principal importance to wholesale dealers. This states:

> "Wholesale distributors must immediately inform the competent authority and the marketing authorisation holder of any medicinal products they identify as falsified or suspect to be falsified[28]. A procedure should be in place to this effect. It should be recorded with all the original details and investigated.
> Any falsified medicinal products found in the supply chain should immediately be physically segregated and stored in a dedicated area away from all other medicinal products. All relevant activities in relation to such products should be documented and records retained."

Wholesale dealers in particular should maintain a high level of vigilance against the procurement or supply of potentially falsified product. Such product may be offered for sale below the established market price so rigorous checks should be made on the bona fides of the supplier and the origin of the product. It is known that some wholesalers are themselves developing good practice strategies – such as conducting rigorous physical inspections of packs when grey market purchases are made – and this is encouraged. Any suspicious activity should be reported to:

[28] 83/83/EC.

Email: casereferrals@mhra.gsi.gov.uk
Telephone: +44 (0)20 3080 6330
To report suspected counterfeit medicines or medical devices:
Email: counterfeit@mhra.gsi.gov.uk
Website: www.mhra.gov.uk
Telephone: +44 (0)20 3080 6701

Regulatory Action

The competent authority will take regulatory action where breaches of legislation are identified; this may take the form of adverse licensing action, e.g. make a variation to an existing licence, suspension or revocation of a licence and/or the instigation of criminal proceedings.

Diverted Medicines

Diversion is the term used for the fraudulent activity where medicines destined for non-EU markets re-enter the EU and are placed back on to the European market at a higher price.

The diversion of medicines involves medicinal products being offered at preferential prices and exported to specific markets (normally third countries) outside the EU. Diversion occurs when unscrupulous traders, on receipt of the medicines, re-export the products back to the EU – with the consequence that patients for whom these preferentially priced medicines were intended, are denied access to them. Such products appearing on the EU market are then known as "diverted" from their intended market. This represents not only a corrupt diversion for profit, but such activity also poses the risk of inappropriate or unlicensed use, and the risk that the product may also be compromised due to poor storage and transportation.

As with counterfeit products, wholesale dealers in particular should maintain a high level of vigilance against the procurement or supply of potentially diverted product. Diverted products may be offered for sale below the established market value, therefore appropriate checks should be made on the bona fides of the supplier and the origin of the product should be ascertained.

Parallel Importation

The UK Parallel Import Licensing Scheme allows nationally authorised medicinal products in other EU Member States to be marketed in the UK, provided the imported products have no therapeutic difference from the equivalent UK products. Parallel importation exists in the absence of price harmonisation of pharmaceutical products within the European Union, i.e. when there are significant price differences between countries; this is the case in the European Union, where prices of medicines are not governed by free competition laws, but are generally fixed by the government.

It involves the transfer of genuine, original branded products, marketed in one Member State of the EEA at a lower price (the source country) to the UK (the country of destination) by a parallel importer, and placed on the market in competition with a therapeutically identical product already marketed there at a higher price by or under licence from the owner of the brand. The scope of the Parallel Import scheme is limited to nationally authorised products, i.e. those medicinal products that have been granted a marketing authorisation by a competent authority of an EEA Member State. It does not extend to centrally authorised medicinal products, granted a marketing authorisation by the EMA which is valid in all Member States. The transfer between EEA markets of these products is termed "Parallel Distribution". The EMA administers a Parallel Distribution scheme for these types of products. Further information on the EMA parallel distribution scheme can be found on the EMA website and in the section below titled "Parallel Distribution".

Products that are parallel imported from another EEA Member State require a Product Licence for Parallel Import (PLPI) granted by the competent regulatory authority (MHRA in the UK) following extensive checks to ensure that the imported drug is therapeutically the same as the domestic version. Further information on the PLPI licensing procedure can be found on the MHRA website.[29]

Parallel importers operating in the UK require a wholesale dealer's licence.

In addition, parallel importers in the UK involved in repackaging or relabelling of product must hold a manufacturer's licence (MIA) authorising product assembly and will be inspected regularly for compliance with GMP. Alternatively, repackaging/relabelling can be contracted out to another company that already holds such a licence.

Parallel importers are required to have effective recall procedures in place. MHRA has systems in place to receive and investigate reports of packaging and labelling problems with medicines, including parallel imported products.

[29] https://www.gov.uk/guidance/medicines-apply-for-a-parallel-import-licence.

Parallel Distribution

Centrally authorised medicinal products are medicines that have been granted a marketing authorisation by the EMA which is valid in all Member States.

The sourcing of centrally authorised medicines (not in the official language of the destination market) from one EEA market to another independent of the Marketing Authorisation Holder is termed Parallel Distribution.

The EMA administers a Parallel Distribution scheme whereby holders of a wholesale dealer's licence wishing to import from other EEA markets must notify the EMA of their intent to import, repackage and distribute the product and provide the EMA with the latest product information and labelling in the language of the Member State of destination to be checked for compliance with the marketing authorisation and latest EU legislation on medicinal products. This requirement is set out in Article 76 of Directive 2001/83/EC on the Community code relating to medicinal products for human use.

Similar to parallel importation, a company wishing to act as a parallel distributor must hold a wholesale dealer's licence. In addition if they intend to carry out the repackaging/relabelling themselves, the company will require an MIA or they can contract this activity out to an MIA holder authorised for assembly.

Further information on Parallel Distribution can be found on the EMA website.[30]

Continued Supply

Under Article 23a of Directive 2001/83/EC, as inserted by Article 1(22) of Directive 2004/27/EC, the marketing authorisation holder is required to notify the competent authority (MHRA in the UK) of the date of actual marketing of the medicinal product, taking account of the various presentations authorised, and to notify the competent authority if the product ceases to be placed on the market either temporarily or permanently. Except in exceptional circumstances, the notification must be made no less than two months before the interruption.

Any authorisation which within three years of granting is not placed on the market will cease to be valid. In respect of generic medicinal products, the three-year period will start on the grant of the authorisation, or at the end of the period of market exclusivity or patent protection of the reference

[30] http://www.ema.europa.eu/ema/index.jsp?curl=pages/regulation/general/general_content_000067.jsp&mid=WC0b01ac0580024594.

product, whichever is the later date. If a product is placed on the market after authorisation, but subsequently ceases to be available on the market in the UK for a period of three consecutive years, it will also cease to be valid. In these circumstances MHRA will, however, when it is aware of the imminent expiry of the three year period, notify the marketing authorisation holder in advance that their marketing authorisation will cease to be valid. In exceptional circumstances, and on public health grounds, MHRA may grant an exemption from the invalidation of the marketing authorisation after three years. Whether there are exceptional circumstances and public health grounds for an exemption will be assessed on a case-by-case basis. When assessing such cases, MHRA will, in particular, consider the implications for patients and public health more generally of a marketing authorisation no longer being valid.

MHRA has received requests for advice on implications for maintaining the harmonisation of an authorisation across Member States if a presentation of a product is withdrawn from the market of the Reference Member State (RMS) and remains unavailable on that market for three years. Discussions on applying the sunset clause provision in such circumstances continue at EU level. In the meantime the MHRA will address the implications of this issue on a case-by-case basis.

Those provisions are implemented in the UK by Part 5 of the Human Medicines Regulations 2012.

In accordance with MHRA's interpretation of the expression "placing on the market" when used elsewhere in the Directive, MHRA's view is that a product is "placed on the market" at the first transaction by which the product enters the distribution chain in the UK. The marketing authorisation holder must, therefore, notify MHRA when a product with a new marketing authorisation is first placed into the distribution chain, rather than the first date it becomes available to individual patients. MHRA requests that you notify us of this first "placing on the market" within one calendar month. In order to ensure that a marketing authorisation continues to be valid, the marketing authorisation holder must ensure that at least one packaging presentation (e.g. bottle or blister pack) of the product, which can include own label supplies, authorised under that marketing authorisation is present on the market.

The marketing authorisation holder must report all cessations/ interruptions to MHRA. However, MHRA does not need to be notified of the following:

(a) normal seasonal changes in manufacturing and/or distribution schedules (such as cold and flu remedies);
(b) short-term temporary interruptions in placing on the market that will not affect normal availability to distributors.

If you are in doubt about whether or not you need to notify an interruption in supply, you should err on the side of caution and report it to MHRA in the normal way. You must notify MHRA if any of the presentations authorised under a single marketing authorisation cease to be placed on the market either temporarily or permanently, but, as stated above, the absence of availability of one or more presentations – as long as one presentation of the product authorised under the single marketing authorisation remains on the market – will not invalidate the marketing authorisation. Problems relating to manufacturing or assembly should also be discussed with the appropriate GMP Inspector and issues of availability of medicines relating to suspected or confirmed product defects should be directly notified to, and discussed with, the Defective Medicines Reporting Centre (Tel: 020 3080 6574).

The Department of Health (DH) also has an interest in the availability of products for supply to the NHS and, together with the Association of the British Pharmaceutical Industry (ABPI) and the British Generics Manufacturers Association (BGMA), has developed best practice guidelines for notifying medicine shortages. These guidelines, together with DH/ABPI guidelines "Ensuring Best Practice in the Notification of Product Discontinuations", complement the statutory requirements under the European legislation and may be found (in pdf format) on the DH website (www.dh.gov.uk). Marketing authorisation holders should, therefore, continue to notify the Department of Health about interruptions and cessations of marketing in accordance with these guidelines.

In this context, your attention is also drawn to Article 81 of Directive 2001/83/EC as substituted by Article 1(57) of Directive 2004/27/EC, under which the marketing authorisation holder and the distributors of a medicinal product actually placed on the market shall, within the limits of their responsibilities, ensure appropriate and continued supplies of that medicinal product to pharmacies and persons authorised to supply medicinal products so that the needs of patients in the Member State in question are covered. Failure by a marketing authorisation holder to comply with this obligation is a criminal offence, unless the marketing authorisation holder took all reasonable precautions and exercised all due diligence to avoid such a failure.

Matters Relating to Unlicensed Medicines

Unless exempt, a medicinal product must be the subject of a marketing authorisation before being placed on the market.

Regulation 167 of the Human Medicines Regulations 2012 provides an exemption from the need for a marketing authorisation for a medicinal product which is supplied:

- in response to an unsolicited order;
- manufactured and assembled in accordance with the specification of a person who is a doctor, dentist, nurse independent prescriber, pharmacist independent prescriber or supplementary prescriber;
- for use by a patient for whose treatment that person is directly responsible in order to fulfil the special needs of that patient; and meets the conditions specified in Regulation 167(2)–(8).

In the interest of public health the exemption is narrowly drawn because these products, unlike licensed medicinal products, may not have been assessed by the Licensing Authority against the criteria of safety, quality and efficacy.

See MHRA Guidance Note 14[31] on the supply of unlicensed medicinal products "specials" which contains additional guidance for those who want to manufacture, import, distribute or supply unlicensed medicines for human use for the treatment of an individual patient.

Importation of unlicensed medicines and centrally authorised products

Unlicensed medicines can be imported into the UK for the special clinical needs of individual patients in accordance with the Human Medicines Regulations [SI 2012/1916], Regulation 167 and the conditions for licence holders (wholesale dealer's licence and manufacturer's "specials" licences) may be found in Schedule 4 of these regulations. More information about importing medicines[32] can be found on the MHRA website.

MHRA regularly receive notifications for the import of unlicensed medicines from importers who have identified shortages of supplies of licensed medicines in the UK. Many of these are quite legitimate, but some have been for import of products that hold centrally authorised marketing authorisations (MAs). These products cannot be imported as unlicensed medicines, as their MAs are valid in all EU Member States. This is the case even if the pack is not the correct UK English language pack and is subject to the European Medicines Agency (EMA) parallel distribution scheme. Import of an unlicensed equivalent would only be permitted if no licensed product (in any pack) could be sourced within the EEA. Importers of unlicensed medicines should ensure that they have procedures in place to

[31] https://www.gov.uk/government/publications/supply-unlicensed-medicinal-products-specials.

[32] https://www.gov.uk/guidance/import-a-human-medicine.

identify centrally authorised medicines as distinct from those which are unlicensed in the UK. The EMA website provides a searchable database of centrally authorised medicines.[33]

The correct method of distribution of centrally authorised medicines is through parallel distribution. The EMA website states that parallel distribution means that a centrally authorised medicine on the market in one Member State is distributed to another Member State by a company independent of the marketing authorisation holder. To be able to sell a medicine in other Member States, parallel distributors need to ensure that the packaging and labelling of the medicine are appropriate, e.g. that the label, box and package leaflet are up to date and available in the correct language. In the UK, if the product is not in the correct pack, this does not make it unlicensed, but the product may be regarded as defective and may lead to regulatory action being taken by MHRA's Defective Medicines Report Centre (DMRC).

Centrally authorised products must always be obtained either from the MA holder or from a registered parallel distributor. The EMA has published an interactive public register of parallel distributors.[34]

Supply of unlicensed medicines when an equivalent licensed product becomes available

Unlicensed medicines may only be supplied against valid special clinical needs of a patient. This requires that there is no authorised equivalent available on the national market. Supply for reasons of cost, institutional need or convenience is not acceptable and is not a special clinical need.

Examples of inappropriate reasons for supply have included preference for a non-parallel imported product, cost, more convenient presentation and longer shelf-life of the unlicensed product. None of these reasons is acceptable.

Whilst requests for procurement of unlicensed medicines are not regulated by MHRA, the supply of unlicensed medicines falls under the Human Medicines Regulations 2012 [SI 2012/1916]. Importers and suppliers must be able to demonstrate compliance with these regulations. This includes supply of unlicensed medicines only to meet valid special clinical needs. Appropriate evidence of supply against such needs should be retained.

[33] http://www.ema.europa.eu/ema/index.jsp?curl=pages/includes/medicines/medicines_landing_page.jsp.

[34] https://fmapps.emea.europa.eu/paradist/index.php.

Reporting Adverse Reactions

An adverse reaction means a response to a medicinal product which is noxious and unintended. A serious adverse reaction is one which results in a person's death, threatens a person's life, results in hospitalisation, persistent or significant disability/incapacity, or results in a congenital abnormality or birth defect.

Wholesalers supplying special medicinal products (unlicensed products) are under an obligation to keep records of any adverse reaction of which they become aware and report any serious adverse reaction to the MHRA; this should be done by submission of a "Yellow Card"[35] report.

Where the product is an exempt advanced therapy medicinal product, the wholesaler is obliged to inform MHRA of any adverse reaction of which they become aware. All records should be retained for the minimum periods required by UK legislation.

Marketing authorisation, homoeopathic registration and traditional herbal medicinal product licence holders have separate obligations in relation to tracking and reporting adverse reactions.

Product Recall/Withdrawal

Manufacturers, importers and distributors are obliged to inform MHRA of any suspected quality defect in a medicinal product that could or would result in a recall, or restriction on supply.

A defective medicinal product is one whose quality does not conform to the requirements of its marketing authorisation, specification or for some other reason of quality is potentially hazardous. A defective product may be suspected because of a visible defect or contamination or as a result of tests performed on it, or because it has caused untoward reactions in a patient or for other reasons involving poor manufacturing or distribution practice. Falsified medicines are considered as defective products.

The Human Medicines Regulations 2012 [SI 2012/1916] imposes certain obligations on licence holders with regard to withdrawal and recall from sale. The aim of the Defective Medicines Report Centre (DMRC) within MHRA is to minimise the hazard to patients arising from the distribution of defective (human) medicinal products by providing an emergency assessment and communications system between the suppliers (manufacturers and distributors), the regulatory authorities and the end user. The DMRC achieves this by receiving reports of suspected defective (human) medicinal products, monitoring and, as far as is necessary,

[35] https://yellowcard.mhra.gov.uk/

directing and advising actions by the relevant licence holder(s) and communicating the details of this action with the appropriate urgency and distribution to users of the products. The communication normally used is a "Drug Alert".

Immediately a hazard is identified from any source, it will be necessary to evaluate the level of danger, and the category of recall, if required. Where the reported defect is a confirmed defect, the DMRC will then take one of the following courses of action and obtain a report from the manufacturer on the nature of the defect, their handling of the defect and action to be taken to prevent its recurrence.

Issue a "Recall"

Under normal circumstances a recall is always required where a defect is confirmed unless the defect is shown to be of a trivial nature and/or there are unlikely to be significant amounts of the affected product remaining in the market.

It is the licence holder's responsibility to recall products from customers, in a manner agreed with the DMRC. The company should provide copies of draft recall letters for agreement with the DMRC. If the company (licence holder) does not agree to a recall voluntarily, MHRA, as licensing authority, may be obliged to take compulsory action.

Issue a "Drug Alert"

Recall and withdrawal of product from the market are normally the responsibility of the licence holder. However, where a product has been distributed widely and/or there is a serious risk to health from the defect, MHRA can opt to issue a Drug Alert letter. The Drug Alert cascade mechanism ensures rapid communication of safety information; it is not a substitute for, but complementary to, any action taken by the licence holder. The text of the Alert should be agreed between MHRA and the company concerned.

In some cases, where a product has been supplied to a small number of known customers, MHRA may decide that notification will be adequate and a Drug Alert is not needed.

The DMRC may also request companies to insert notification in the professional press in certain cases.

Management of the recall

The company should directly contact wholesalers, hospitals, retail pharmacies and overseas distributors supplied. The DMRC is likely to take the lead in notifying Regional Contacts for NHS Trusts and Provider Units and Health Authorities, special and Government hospitals and overseas regulatory authorities.

The DMRC will liaise with the company and discuss arrangements for the recall, requesting the dates that supply started and ceased and a copy of any letters sent out by that company concerning the recall. Again, it is desirable that the text of the notices sent via the company and by the DMRC should be mutually agreed.

Management of recall activity by wholesalers

MHRA expects key personnel, such as the Responsible Person, to keep themselves up to date with drug safety issues. This can be done, for example, by checking the relevant sections of MHRA's website daily or alternatively by signing up for relevant MHRA email alerts.

Wholesalers and brokers should retain records of all recall notices received and/or reviewed, including those notices for which no action is required.

Recall processes should be designed with quality management in mind, for example change control and risk management should be applied for significant changes such as a change of Responsible Person or nominated deputy, changes to product handling/storage, or changes to transport arrangements. At such times there should be an assessment of whether a test recall should be performed to provide assurance of the ongoing effectiveness of the company's processes and identify any weaknesses or areas for improvement.

Chapter 6 section 5 of the EU GDP Guidelines[36] describes the minimum standards for wholesalers in relation to recall activities.

When the wholesaler receives a drug alert or recall notice they should take steps to:

- follow the detail of the recall message;
- identify any affected stock on site or in transit (including that returned from customers);
- identify any customers to whom the affected products have been supplied, including where products may have been supplied as samples;
- directly contact those customers, making them aware of the details of the recall and where necessary providing a mechanism for returning affected stock;

[36] Guidelines of 5 November 2013 on Good Distribution Practice of medicinal products for human use (Directive 2013/C 343/01).

- physically segregate and quarantine any affected product in an area away from other medicinal products, ensuring that such stock does not re-enter the supply chain;
- reconcile the quantities of stock purchased, onsite, sent out and returned;
- keep recall notices open for an appropriate period, so as to capture any affected stock still moving through the supply chain.

All recall activities should be documented at the time they occur and at the conclusion of all recall activities the Responsible Person (or their nominated deputy) should produce a report, making an objective assessment of whether the recall process achieved its objectives and identifying any areas requiring improvement.

Testing the recall process

Section 6.5 of the EU GDP Guidelines requires that "the effectiveness of the arrangements for product recall should be evaluated regularly (at least annually)".

The aim of such an evaluation should be to challenge the internal processes of the licence holder as far as practicable. This is particularly important in organisations where recall activities occur infrequently, for example where the product range handled is limited in quantity or scope. Where stock is held on behalf of a licence holder at a third party site then the recall test should extend to covering activities at the third party site; this will require a degree of liaison between both sites' Responsible Persons.

The test process should be described in the company's quality system in sufficient detail to allow staff to perform the test and to be able to assess the progress of the recall process at each step.

It is expected that as a minimum the test process should mimic a real recall but should stop short of contacting the licence holder's customers. The product and batch selected should be typical of those handled by the company in the previous 12 months and where possible provide a worst case scenario (for example when key staff are absent or for essential medicines when alternative supplies may need to be made available). Correspondence (for example emails to staff at a branch level) should, as far as practicable, not indicate that it is being used for test purposes.

The test process should be documented so as to be able to demonstrate that:

- staff can receive information effectively and act on it quickly;
- stock in question can be identified, reconciled and segregated effectively;
- customers supplied with the stock can be identified quickly from records, and their most up-to-date contact details confirmed (including address and telephone details both inside and outside working hours);

- adherence to the company's recall process has taken place;
- the effectiveness of training can be assessed.

For the purposes of a test recall MHRA does not normally expect licence holders to contact their customers as this could lead to unintended consequences should the customer believe the test recall to be real. Instead, the licence holder should obtain evidence that the contact details (physical address, telephone/fax numbers, email address) they hold for the relevant customers are up to date. Where the licence holder would send a letter/fax/email to their customer with the details of the recall, this should be drafted but not sent.

Where customers are contacted, there should be adequate oversight of the entire test process to ensure that misunderstandings do not occur.

As with all routine recall activities, at the conclusion of test recall activities the Responsible Person should produce a report, making an objective assessment of whether the recall process achieved its objectives and identifying any areas requiring improvement.

It may be possible for a company to use recall activities for non-medicinal products to demonstrate an effective recall process, provided that the process described in the company's quality system covers the handling of both medicinal and non-medicinal products in the same manner and leads to the same outcome.

Follow-up action

The DMRC will monitor the conduct and success of the recall by the manufacturer or distributor. As follow-up action, it may be necessary to consider any or all of the following:

- arrange a visit to the licence holder/manufacturer/distributor;
- arrange a visit to the point of discovery of the defect;
- refer to the Inspectorate to arrange an inspection;
- seek special surveillance of adverse reaction reports;
- refer the matter for adverse licensing and/or enforcement action.

Reporting a suspected defect

Suspected defects can be reported by telephone, email or letter or using our online form:

Address:
DMRC, 151 Buckingham Palace Road, London SW1W 9SZ, UK.
Telephone: +44 (0)20 3080 6574 (08:45–16:45 Monday to Friday)
Telephone: +44 (0)7795 641532 (urgent calls outside working hours, at weekends or on public holidays)
Email: dmrc@mhra.gsi.gov.uk
Online form: www.mhra.gov.uk
http://www.mhra.gov.uk/Safetyinformation/Reportingsafetyproblems/Reportingsuspecteddefectsinmedicines/Suspecteddefectonlineform/index.htm

Data Integrity

Data integrity (DI) is the extent to which all data are complete, consistent and accurate throughout the data lifecycle. It is essential that this principle of DI is followed in order to ensure traceability of GDP activities and associated records and other documentation. Failure to meet this requirement has been present in all areas of GxP, and as a result, the MHRA has provided guidance including definition and concept papers published as Inspectorate blogs[1]. In addition MHRA has published a DI guidance document for consultation by industry[2] and has been involved in the development of DI guidance published by other regulatory authorities. The reader is encouraged to refer to these documents for a complete list of definitions and broader guidance. This article aims to provide guidance specific to implementation of DI within the GDP environment.

Basic concepts

Data Integrity is not a new regulatory requirement but has been a common failure present in multiple quality system components and applies to both electronic and hard copy data. There are no specific inspection deficiencies for DI failure within MHRA GDP classification; however where a DI failure is observed it is described in relation to the EU GDP Guidelines.

[1] Good Manufacturing Practice (GMP) data integrity: a new look at an old topic, part 1, David Churchward, 25 June 2015, MHRA Inspectorate blog.

[2] MHRA data integrity guidance: 18 months on, David Churchward, 21 July 2016, MHRA Inspectorate blog.

An example may be failure to ensure that data was readily available or retrievable (GDP Guidelines chapter 4.2) where original hard copy data had been obscured with correction fluid.

Failure in DI is often caused by weakness in implementation of measures that ensure that DI standards are implemented and maintained, referred to as Data Governance. The MHRA expects that appropriate measures are taken in respect to DI and Data Governance, and it should be clear which data is critical to regulatory compliance and product safety in order to ensure appropriate resource is applied to the more critical data. For data which is not as important, less effort will be expected along with a rationale as to what data is deemed by the company as being not critical.

Criticality of data

Quality risk management lends itself as a useful tool to determine which data and records are critical to operations and therefore require more robust data governance measures. If GDP activity cannot be reconstructed should specific data be lost, then it can be assumed that the data in question is critical. For equipment that generates data the extent of qualification in respect to DI should correlate with data criticality. Quality risk management can also be applied to identify which data and records are not critical such as secondary records and put in place measures to ensure that critical data does not inadvertently end up on the secondary record. An example is of a customer invoice which gets copied on receipt of stock and retained in the warehouse as evidence of delivery while the original is sent to the accountant. If the warehouse copy is annotated, e.g. to amend stock quantity booked in, then it is no longer a true copy of the original and therefore both need to be considered as master documents.

If data is transcribed from an instrument or hard copy record onto a computerised system and the original record is considered by the company to be non-critical, then consideration must be made as to how the company considers the original record as not being critical. Appropriate measures of review and approval of the transcribed records must be put in place. Any inconsistency in the computerised record would indicate failure not only in the integrity of the data but also failure of the governance process concerned with the review and approval of transcription, and any other records approved in the same way including records not related to this event would be circumspect.

ALCOA applied to GDP

A - ATTRIBUTABLE TO THE PERSON GENERATING THE DATA

Any critical data or information recorded for GDP purposes must be attributed to the originator. For hard copy records, entries should be traceable to the person making the record, with initialling and dating being developed as an unconscious habit. Where computerised systems are in use individuals are expected to only access systems by a unique password that can be traced to a level of permission. To guarantee this there should be adequate provision of terminals and a culture of logging out when not using shared devices to prevent work-around arrangements from being developed. A person with administrator access must not use this mode for anything other than maintenance operations and should have a separate user account for daily operations.

Where signatures or names relate to personnel outside of the organisation, e.g. engineer, then the name should be printed in addition to signing and they should also print their job role. For regular contacts, e.g. customers, a signature log may be developed and managed as a controlled document. Where shipments are international and parts of the delivery are outsourced, additional control measures should be put in place to maintain the chain of custody for the delivery from the warehouse to the customer.

L - LEGIBLE AND PERMANENT

Hard copy records are often not legible due to poor training of staff, poor form design and poor process design and lead to errors in reading and transcription. Simple solutions include provision of clipboards to prevent warehouse and delivery forms being completed on top of non-flat surfaces leading to poor writing, development of forms that accommodate those with large writing, and training staff in how to record errors and how to use traceable footnotes rather than squeeze notes into a small comments box. Replacement of hard copy with electronic records may improve legibility but may pose other problems such as poor accessibility to records.

The trend towards replacing manual systems with computerised ones has not always been met with appropriate assessment of risk, for example, there may be emphasis placed on data back-up of records but little attention paid to ability to restore records. Where computer systems are updated and previous software or hardware is no longer supported then a quality risk assessment should be carried out and appropriate action undertaken to ensure records can be retrieved.

The versatility and ease of use of electronic spreadsheets has led to them being very common. They lose the ability to retain original data that is overwritten and entries are normally not attributable to the recorder unless strict document control measures are put in place, including access control and versioning. Other approaches to control spreadsheets include printing in hard copy or pdf form and retaining a log of each approved version.

C - CONTEMPORANEOUS

Records should be traceable to the time the activity is carried out in order to reduce the chance of the record being forgotten or traceability of actions lost. Some events are more time-critical than others, such as execution of a medicine recall or qualification of a transport lane. In these circumstances recording of events must be consistent especially where activity spans different time zones or different date formats are in use.

O - ORIGINAL RECORD (OR TRUE COPY)

The original record refers to data as originally generated, preserving the integrity (accuracy, completeness, content and meaning) of the record, e.g. original paper record of manual observation, or electronic raw data file from a computerised system.

A true copy refers to a copy of original information that been verified as an exact (accurate and complete) copy having all of the same attributes and information as the original. The copy may be verified by dated signature or by a validated electronic signature.

Where hard copy documents are scanned into electronic format for archiving there needs to be a process of verification that all records are complete and are an accurate representation of the original. Problems can exist where some documents in a bundle are of poor print quality, double-sided or where a highlighter has been used. In these cases the electronic records may not be complete or a true copy of the original.

On occasions where documents are provided from third parties, such as copies of or translations of customer wholesale dealer's licences, the document must be authenticated as a true copy by reference to an appropriate source such as a regulatory authority. Reliance solely on the word of the third party is not acceptable.

Where copies of originals are made they should be clearly able to be differentiated from the original and prevent mix-up. Possible control measures include use of watermarks, embossing or having original documents on coloured paper with restricted access to the paper.

A - ACCURATE

Accuracy of data is essential to GDP, and having good processes to manage errors supports this. Deviation management should ensure corrections are traceable and approved, and original incorrect data not lost. Processes and systems should be developed that drive accuracy rather than challenge it and where data is manipulated then there should be defined rules controlling this e.g. number rounding and conversion of units of measure. A common failure in this respect is in management of stock adjustments where physical stock count and stock records do not match. Stock records need to be managed in an open and honest manner and adjustments not hidden but corrected with appropriate justification and authorisation.

When formulae are used in electronic spreadsheets they are rarely qualified, in which case errors can be introduced without being noticed. The use of check boxes and formulae that detect nonsense values can help reduce errors.

Data Governance

Having good quality data and records is not only essential to GDP but also a contributory factor in managing an effective and efficient operation. A fundamental requirement is development of a quality culture where all staff are able to identify weaknesses without feeling intimidated and understand the importance of maintaining accurate records and adherence to procedures.

Good training and level of knowledge is also required, especially for staff in quality assurance roles as they are often responsible for provision of training, design of processes and procedures, evaluation of deviations and creation of quality culture. If they are weak in any of these then the staff required to make accurate records are at a disadvantage. Training can also be provided specifically in relation to DI so, for example, staff understand the difference between a witness signature and a check signature and the risks to DI associated with a particular process, as well as reviewing records to ensure that data makes sense in addition to confirming all entries are complete.

One of the most common failures in GDP is inadequate control of quality system documents. Where events such as complaints or deviations are recorded in free-vend template forms these are often not reconciled or reconcilable. This may lead to records being lost or incorrect template versions being used. Good system design with consideration of DI provides the means to ensure all records are complete with use of simple solutions, such as use of hard-bound forms or controlled issue of numbered and indexed forms.

Data integrity can be monitored by incorporation into self-inspections or as a single separate horizontal audit to enable best practice to be shared across departments. Monitoring of near misses in addition to full breaches and consideration of opportunities for continuous improvement all add to the data governance tools that can be reviewed during quality management reviews that in turn lead to further development of a healthy quality culture.

Brokering of Medicines

CHAPTER

6

EU Legislation on Brokering Medicines

Contents

DIRECTIVE 2001/83/EC, AS AMENDED, TITLE VII, WHOLESALE DISTRIBUTION AND BROKERING MEDICINES

Directive 2001/83/EC of the European Parliament and of the Council of 6 November 2001 on the Community code relating to medicinal products for human use as amended

Editor's note The Articles reproduced below are those relevant to persons who broker medicines. For the full text of Title VII of Directive 2001/83/EC, please refer to Chapter 2 earlier in this guide.

Title VII: Wholesale Distribution and Brokering of Medicinal Products

Article 80

Holders of the distribution authorization must fulfil the following minimum requirements:

(a) they must make the premises, installations and equipment referred to in Article 79(a) accessible at all times to the persons responsible for inspecting them;

(**b**) they must obtain their supplies of medicinal products only from persons who are themselves in possession of the distribution authorization or who are exempt from obtaining such authorization under the terms of Article 77(3);

(**c**) they must supply medicinal products only to persons who are themselves in possession of the distribution authorization or who are authorized or entitled to supply medicinal products to the public in the Member State concerned;

(**ca**) they must verify that the medicinal products received are not falsified by checking the safety features on the outer packaging, in accordance with the requirements laid down in the delegated acts referred to in Article 54a(2);

(**d**) they must have an emergency plan which ensures effective implementation of any recall from the market ordered by the competent authorities or carried out in cooperation with the manufacturer or marketing authorization holder for the medicinal product concerned;

(**e**) they must keep records either in the form of purchase/sales invoices or on computer, or in any other form, giving for any transaction in medicinal products received, dispatched or brokered at least the following information:

- date,
- name of the medicinal product,
- quantity received, supplied or brokered,
- name and address of the supplier or consignee, as appropriate,
- batch number of the medicinal products at least for products bearing the safety features referred to in point (o) of Article 54;

(**f**) they must keep the records referred to under (e) available to the competent authorities, for inspection purposes, for a period of five years;

(**g**) they must comply with the principles and guidelines of good distribution practice for medicinal products as laid down in Article 84.

(**h**) they must maintain a quality system setting out responsibilities, processes and risk management measures in relation to their activities;

(**i**) they must immediately inform the competent authority and, where applicable, the marketing authorisation holder, of medicinal products they receive or are offered which they identify as falsified or suspect to be falsified.

For the purposes of point (b), where the medicinal product is obtained from another wholesale distributor, wholesale distribution authorisation holders must verify compliance with the principles and guidelines of good distribution practices by the supplying wholesale distributor. This includes verifying whether the supplying wholesale distributor holds a wholesale distribution authorisation.

Where the medicinal product is obtained from the manufacturer or importer, wholesale distribution authorisation holders must verify that the manufacturer or importer holds a manufacturing authorisation.

Where the medicinal product is obtained through brokering, the wholesale distribution authorisation holders must verify that the broker involved fulfils the requirements set out in this Directive.

Article 85b

1 Persons brokering medicinal products shall ensure that the brokered medicinal products are covered by a marketing authorisation granted pursuant to Regulation (EC) No 726/2004 or by the competent authorities of a Member State in accordance with this Directive.

Persons brokering medicinal products shall have a permanent address and contact details in the Union, so as to ensure accurate identification, location, communication and supervision of their activities by competent authorities.

The requirements set out in points (d) to (i) of Article 80 shall apply *mutatis mutandis* to the brokering of medicinal products.

2 Persons may only broker medicinal products if they are registered with the competent authority of the Member State of their permanent address referred to in paragraph 1. Those persons shall submit, at least, their name, corporate name and permanent address in order to register. They shall notify the competent authority of any changes thereof without unnecessary delay.

Persons brokering medicinal products who had commenced their activity before 2 January 2013 shall register with the competent authority by 2 March 2013.

The competent authority shall enter the information referred to in the first subparagraph in a register that shall be publicly accessible.

3 The guidelines referred to in Article 84 shall include specific provisions for brokering.

4 This Article shall be without prejudice to Article 111. Inspections referred to in Article 111 shall be carried out under the responsibility of the Member State where the person brokering medicinal products is registered.

If a person brokering medicinal products does not comply with the requirements set out in this Article, the competent authority may decide to remove that person from the register referred to in paragraph 2. The competent authority shall notify that person thereof.

CHAPTER

7

UK Legislation on Brokering Medicines

Contents

The Human Medicines Regulations 2012 [SI 2012/1916]

Editor's note These extracts from the Human Medicines Regulations 2012 [SI 2012/1916] as amended are presented for the reader's convenience. Reproduction is with the permission of HMSO and the Queen's Printer for Scotland. For any definitive information reference must be made to the original amending Regulations. The numbering and content within this section correspond with the regulations set out in the published Statutory Instrument [SI 2012/1916] as amended.

Citation and commencement

1 (1) These Regulations may be cited as the Human Medicines Regulations 2012.

(2) These Regulations come into force on 14th August 2012.

General interpretation

8 (1) In these Regulations (unless the context otherwise requires)-
"brokering" means all activities in relation to the sale or purchase of medicinal products, except for wholesale distribution, that do not

include physical handling and that consist of negotiating independently and on behalf of another legal or natural person;
"falsified medicinal product" means any medicinal product with a false representation of:

(a) its identity, including its packaging and labelling, its name or its composition (other than any unintentional quality defect) as regards any of its ingredients including excipients and the strength of those ingredients;
(b) its source, including its manufacturer, its country of manufacturing, its country of origin or its marketing authorisation holder; or
(c) its history, including the records and documents relating to the distribution channels used.

Brokering in medicinal products

45A. (1) A person may not broker a medicinal product unless:
(a) that product is covered by an authorisation granted:
(i) under Regulation (EC) No 726/2004; or
(ii) by a competent authority of a member State; and
(b) that person:
(i) is validly registered as a broker with a competent authority of a member State,
(ii) except where the person is validly registered with the competent authority of another EEA state, has a permanent address in the United Kingdom, and
(iii) complies with the guidelines on good distribution practice published by the European Commission in accordance with Article 84 of the 2001 Directive insofar as those guidelines apply to brokers.

(2) A person is not validly registered for the purpose of paragraph (1)(b) if:
(d) the person's permanent address is not entered into a register of brokers kept by a competent authority of a member State;
(e) the registration is suspended; or
(f) the person has notified the competent authority of a member State to remove that person from the register.

(3) Paragraph (1)(b)(i) does not apply until 20th October 2013 in relation to a person who brokered any medicinal product before 20th August 2013.

Application for brokering registration

45B. (1) The licensing authority may not register a person as a broker unless paragraphs (2) to (7) are complied with.

(2) An application for registration must be made containing:

(a) the name of the person to be registered;

(b) the name under which that person is trading (if different to the name of that person);

(c) that person's:

(i) permanent address in the United Kingdom,

(ii) e-mail address, and

(iii) telephone number;

(d) a statement of whether the medicinal products to be brokered are:

(i) prescription only medicines,

(ii) pharmacy medicines, or

(iii) medicines subject to general sale;

(e) an indication of the range of medicinal products to be brokered;

(f) evidence that that person can comply with regulations 45A(1)(b)(iii), 45E(3)(a) to (f) and 45F(1); and

(g) any fee payable in connection with the application in accordance with the Fees Regulations.

(3) Where the address at which the emergency plan, documents or record necessary to comply with regulation 45E(3)(b) to (d) are kept is different from the address notified in accordance with sub-paragraph (2)(c)(i), the application must contain:

(a) that address where the plan or records are to be kept;

(b) the name of a person who can provide access to that address for the purpose of regulation 325 (rights of entry); and

(c) that person's:

(i) address,

(ii) e-mail address, and

(iii) telephone number.

(4) Unless paragraph (6) applies, the application for registration must:

(a) be in English; and

(b) be signed by the person seeking a brokering registration.

(5) The pages of the application must be serially numbered.

(6) Where the application is made on behalf of the person seeking a brokering registration by another person ("A"), the application must:

(a) contain the name and address of A; and

(b) be signed by A.

Criteria of broker's registration

45E. (1) Registration of a broker is conditional on that broker:
(a) complying with regulation 45A(1); and
(b) satisfying:
(i) the criteria in paragraphs (3), (4) and (7), and
(ii) such other criteria as the licensing authority considers appropriate and notifies the broker of.
(2) The criteria referred to in paragraph (1)(b)(ii) may include (but are not limited to) the criteria specified in paragraphs (5) and (6).
(3) The broker must:
(a) have a permanent address in the United Kingdom;
(b) maintain an emergency plan to ensure effective implementation of the recall from the market of a medicinal product where recall is:
(i) ordered by the licensing authority or by the competent authority of any EEA State, or
(ii) carried out in co-operation with the manufacturer of, or the holder of the marketing authorisation, for the product;
(c) keep documents relating to the sale or supply of medicinal products under the licence which may facilitate the withdrawal or recall from sale of medicinal products in accordance with sub-paragraph (b);
(d) record in relation to the brokering of each medicinal product:
(i) the name of the medicinal product,
(ii) the quantity of the product brokered,
(iii) the batch number of the medicinal product bearing the safety features referred to in point (o) of Article 54 of the 2001 Directive,
(iv) the name and address of the:
(aa) supplier, or
(bb) consignee, and
(v) the date on which the sale or purchase of the product is brokered;
(e) maintain a quality system setting out responsibilities, processes and risk management measures in relation to their activities; and
(f) keep the documents or record required by sub-paragraph (c) or (d) available to the licensing authority for a period of five years; and
(g) comply with regulation 45F(1), (2) and (4).
(4) Where the address at which the plan or records necessary to comply with paragraph (3)(b) to (d) are kept is different from the address notified in accordance with regulation 45B(2)(c)(i), the broker must:
(a) ensure that the plan or records are kept at an address in the United Kingdom; and

(b) inform the licensing authority of the address at which the plan or records are kept.

(5) The broker must provide such information as may be requested by the licensing authority concerning the type and quantity of medicinal products brokered within the period specified by the licensing authority.

(6) The broker must take all reasonable precautions and exercise all due diligence to ensure that any information provided by that broker to the licensing authority in accordance with regulation 45F is not false or misleading.

(7) For the purposes of enabling the licensing authority to determine whether there are grounds for suspending, revoking or varying the registration, the broker must permit a person authorised in writing by the licensing authority, on production of identification, to carry out any inspection, or to take any copies, which an inspector may carry out or take under regulations 325 (rights of entry) and 327 (powers of inspection, sampling and seizure).

Provision of information

45F. (1) A broker registered in the UK must immediately inform:

(a) the licensing authority; and

(b) where applicable, the marketing authorisation holder, of medicinal products which the broker identifies as, suspects to be, or has reasonable grounds for knowing or suspecting to be, falsified.

(2) On or before the date specified in paragraph (3), a broker who is, or has applied to the licensing authority to become, a registered broker in the United Kingdom must submit a report to the licensing authority, which:

(a) includes a declaration that the broker has in place an appropriate system to ensure compliance with regulations 45A, 45B and this regulation; and

(b) details the system which the broker has in place to ensure such compliance.

(3) The date specified for the purposes of this paragraph is:

(a) in relation to any application made before 31st March 2014, the date of the application; and

(b) in relation to each subsequent reporting year, 30th April following the end of that year.

(4) The broker must without delay notify the licensing authority of any changes to the matters in respect of which evidence has been supplied in relation to paragraph (2) which might affect compliance with the requirements of this Chapter.

(5) Any report or notification to the licensing authority under paragraph (2) or (4) must be accompanied by the appropriate fee in accordance with the Fees Regulations.
(6) The licensing authority may give a notice to a registered broker requiring that broker to provide information of a kind specified in the notice within the period specified in the notice.
(7) A notice under paragraph (6) may not be given to a registered broker unless it appears to the licensing authority that it is necessary for the licensing authority to consider whether the registration should be varied, suspended or revoked.
(8) A notice under paragraph (6) may specify information which the licensing authority thinks necessary for considering whether the registration should be varied, suspended or revoked.
(9) In paragraph (3)(b), "reporting year" means a period of twelve months ending on 31st March.

CHAPTER

8

EU Guidelines on Good Distribution Practice of Medicinal Products for Human Use (2013/C 343/01)

Contents

Guidelines on Good Distribution Practice of Medicinal Products for Human Use (2013/C 343/01)

Editor's note: The chapters reproduced below are those relevant to persons that broker medicines. For the full text of the Guidelines on Good Distribution Practice of Medicinal Products for Human Use (2013/C 343/01), please refer to Chapter 4 earlier in this guide.

Chapter 4 – Documentation

4.1 Principle

Good documentation constitutes an essential part of the quality system. Written documentation should prevent errors from spoken communication and permits the tracking of relevant operations during the distribution of medicinal products.

4.2 General

Documentation comprises all written procedures, instructions, contracts, records and data, in paper or in electronic form. Documentation should be readily available/retrievable.

With regard to the processing of personal data of employees, complainants or any other natural person, Directive 95/46/EC[1] on the protection of individuals applies to the processing of personal data and to the free movement of such data.

Documentation should be sufficiently comprehensive with respect to the scope of the wholesale distributor's activities and in a language understood by personnel. It should be written in clear, unambiguous language and be free from errors.

Procedure should be approved, signed and dated by the responsible person. Documentation should be approved, signed and dated by appropriate authorised persons, as required. It should not be handwritten; although, where it is necessary, sufficient space should be provided for such entries.

Any alteration made in the documentation should be signed and dated; the alteration should permit the reading of the original information. Where appropriate, the reason for the alteration should be recorded.

Documents should be retained for the period stated in national legislation but at least five years. Personal data should be deleted or anonymised as soon as their storage is no longer than necessary for the purpose of distribution activities.

Each employee should have ready access to all necessary documentation for the tasks executed.

Attention should be paid to using valid and approved procedures. Documents should have unambiguous content; title, nature and purpose should be clearly stated. Documents should be reviewed regularly and kept up to date. Version control should be applied to procedures. After revision of a document a system should exist to prevent inadvertent use of the superseded version. Superseded or obsolete procedures should be removed from workstations and archived.

Records must be kept either in the form of purchase/sales invoices, delivery slips, or on computer or any other form, for any transaction in medicinal products received, supplied or brokered.

[1] OJ L 281, 23.11.1995, p. 31.

Records must include at least the following information: date; name of the medicinal product; quantity received, supplied or brokered; name and address of the supplier, customer, broker or consignee, as appropriate; and batch number at least for medicinal product bearing the safety features[2].

Records should be made at the time each operation is undertaken.

Chapter 10 – Specific Provisions for Brokers[3]

10.1 Principle

A 'broker' is a person involved in activities in relation to the sale or purchase of medicinal products, except for wholesale distribution, that do not include physical handling and that consist of negotiating independently and on behalf of another legal or natural person.[4]

Brokers are subject to a registration requirement. They must have a permanent address and contact details in the Member State where they are registered[5]. They must notify the competent authority of any changes to those details without unnecessary delay.

By definition, brokers do not procure, supply or hold medicines. Therefore, requirements for premises, installations and equipment as set out in Directive 2001/83/EC do not apply. However, all other rules in Directive 2001/83/EC that apply to wholesale distributors also apply to brokers.

10.2 Quality system

The quality system of a broker should be defined in writing, approved and kept up to date. It should set out responsibilities, processes and risk management in relation to their activities.

The quality system should include an emergency plan which ensures effective recall of medicinal products from the market ordered by the manufacturer or the competent authorities or carried out in cooperation with the manufacturer or marketing authorisation holder for the medicinal product concerned[6]. The competent authorities must be immediately informed of any suspected falsified medicines offered in the supply chain[7].

[2] Articles 80(e) and 82 of Directive 2001/83/EC.
[3] Article 1(17a) of Directive 2001/83/EC.
[4] Article 85b of Directive 2001/83/EC.
[5] Article 80(d) of Directive 2001/83/EC.
[6] Article 85b(1), third paragraph of Directive 2001/83/EC.
[7] Article 85b(3) of Directive 2001/83/EC.

10.3 Personnel

Any member of personnel involved in the brokering activities should be trained in the applicable EU and national legislation and in the issues concerning falsified medicinal products.

10.4 Documentation

The general provisions on documentation in Chapter 4 apply.

In addition, at least the following procedures and instructions, along with the corresponding records of execution, should be in place:

(i) procedure for complaints handling;
(ii) procedure for informing competent authorities and marketing authorisation holders of suspected falsified medicinal products;
(iii) procedure for supporting recalls;
(iv) procedure for ensuring that medicinal products brokered have a marketing authorisation;
(v) procedure for verifying that their supplying wholesale distributors hold a distribution authorisation, their supplying manufacturers or importers hold a manufacturing authorisation and their customers are authorised to supply medicinal products in the Member State concerned;
(vi) records should be kept either in the form of purchase/sales invoices or on computer, or in any other form for any transaction in medicinal products brokered and should contain at least the following information: date; name of the medicinal product; quantity brokered; name and address of the supplier and the customer; and batch number at least for products bearing the safety features.

Records should be made available to the competent authorities, for inspection purposes, for the period stated in national legislation but at least five years.

CHAPTER

9

UK Guidance on Brokering Medicines

Contents

Introduction 206
Brokering in Medicinal Products 206
Registration 207
Application for Brokering Registration 207
Criteria of Broker's Registration 208
Provision of Information 209
Good Distribution Practice 210
Management of Recall Activity by Brokers 210
Supply and brokering to countries outside of the UK 211

Introduction

Persons procuring, holding, storing, supplying or exporting medicinal products are required to hold a wholesale distribution authorisation in accordance with Directive 2001/83/EC which lays down the rules for the wholesale distribution of medicinal products in the Union.

However, the distribution network for medicinal products may involve operators who are not necessarily authorised wholesale distributors. To ensure the reliability of the supply chain, Directive 2011/62/EU, the Falsified Medicines Directive extends medicine legislation to the entire supply chain. This now includes not only wholesale distributors, whether or not they physically handle the medicinal products, but also brokers who are involved in the sale or purchase of medicinal products without selling or purchasing those products themselves, and without owning and physically handling the medicinal products.

Brokering in Medicinal Products

Brokering of medicinal products is defined in the Falsified Medicines Directive and means:

All activities in relation to the sale or purchase of medicinal products, except for wholesale distribution, that do not include physical handling and that consist of negotiating independently and on behalf of another legal or natural person.

To accord with the Directive brokers may only broker medicinal products that are the subject of an authorisation granted by the European Commission or a National Competent Authority.

Brokers should be established at a permanent address and have contact details in the EU and may only operate following registration of these details with the National Competent Authority. In the UK this is MHRA. A broker must provide required details for registration which will include their name, corporate name and permanent address. They must also notify any changes without unnecessary delay. This is to ensure the brokers accurate identification, location, communication and supervision of their activities by the National Competent Authorities.

Brokers can negotiate between the manufacturer and a wholesaler, or one wholesaler and another wholesaler, or the manufacturer or wholesale dealer with a person who may lawfully sell those products by retail or may lawfully supply them in circumstances corresponding to retail sale or a person who may lawfully administer those products.

Brokers are not virtual wholesale dealers; the definition of "Brokering medicinal products" specifically excludes the activity of "wholesale dealing". Wholesale dealing and brokering of medicinal products are separate activities. Therefore wholesale dealers who wish to broker will require a separate registration, because:

- EU legislation defines wholesale distribution and brokering a medicinal product separately;
- wholesale dealers are licensed;
- the brokering of medicinal products is subject to registration.

Registration

UK based companies that broker medicinal products and are involved in the sale or purchase of medicinal products without selling or purchasing those products themselves, and without owning and physically handling the medicinal products are considered to be brokers and will have to register with MHRA.

In order to register in the UK, brokers will have a permanent address and contact details in the UK and will only be allowed to operate as a bona fide broker following their successful registration with MHRA.

Application for Brokering Registration

The registration regime for UK brokers is subject to an application procedure, followed by a determination procedure completed by MHRA. The application procedure will include:

- making an application for registration;
- assessment by MHRA of the application;
- providing specific evidence to check bona fides;
- advising an applicant of the decision.

UK brokers may be subject to inspection at their registered premises. This will be under a risk-based inspection programme. Once registered a broker's registration will be recognised by other Member States and will allow the broker to broker across the EEA. UK medicines legislation in respect of brokering will also recognise registered brokers in other EEA Member States in the same way.

MHRA has an obligation to enter the information on a publicly accessible UK register following the determination of successful application for registration.

This publicly available UK register is required to enable National Competent Authorities in other EEA Member States to establish the bona fides and compliance of brokers established in the UK where they are involved in the sale or purchase of medicines on their territories and the UK will investigate complaints of non-compliance. Reciprocal arrangements will apply for brokers established in other Member States involved in the sale or purchase of medicines to and from the UK.

Criteria of Broker's Registration

A person may not broker a medicinal product unless that product is covered by an authorisation granted under Regulation (EC) No 726/2004 or by a competent authority of an EEA Member State and that person is validly registered as a broker with a competent authority of an EEA Member State.

A broker is not validly registered if the broker's permanent address is not entered into a register of brokers kept by a competent authority of a Member State or the registration is suspended or the broker has notified the competent authority of an EEA Member State to remove them from the register.

Brokers must satisfy all the conditions of brokering and:

- have a permanent address in the UK;
- have an emergency plan which ensures effective implementation of any recall from the market ordered by the competent authorities or carried out in cooperation with the manufacturer or marketing authorisation holder for the medicinal product concerned;
- keep records either in the form of purchase/sales invoices or on computer, or in any other form, giving for any transaction in medicinal products brokered at least the following information:

- date on which the sale or purchase of the product is brokered;
- name of the medicinal product;
- quantity brokered;
- name and address of the supplier or consignee, as appropriate;
- batch number of the medicinal products at least for products bearing the safety features referred to in point (o) of Article 54 of Directive 2001/83/EC;

- keep the records available to the competent authorities, for inspection purposes, for a period of five years;
- comply with the principles and guidelines of good distribution practice for medicinal products as laid down in Article 84 of Directive 2001/83/EC;
- maintain a quality system setting out responsibilities, processes and risk management measures in relation to their activities.

Where the address at which the plan or records necessary to comply with the provisions of brokering are kept is different from the address notified in accordance with the application, the broker must ensure that the plan or records are kept at an address in the UK and inform the licensing authority of the address at which the plan or records are kept.

The broker must provide such information as may be requested by MHRA concerning the type and quantity of medicinal products brokered within the period specified by MHRA.

The broker must take all reasonable precautions and exercise all due diligence to ensure that any information provided by that broker to MHRA is not false or misleading.

For the purposes of enabling MHRA to determine whether there are grounds for suspending, revoking or varying the registration, the broker must permit a person authorised in writing by MHRA, on production of identification, to carry out any inspection, or to take any copies, which an inspector may carry out or take under the provisions of the Human Medicines Regulations 2012 [SI 2012/1916].

Provision of Information

Once registered, a broker will have to notify MHRA of any changes to the details for registration which might affect compliance with the requirements of the legislation in respect of brokering without unnecessary delay. This notification will be subject to a variation procedure. Responsibility for notifying MHRA of any changes lies with the person responsible for management of the brokering activities.

The person responsible for management of the brokering activities shall be required to submit a report which shall include:

- a declaration that the broker has in place appropriate systems to ensure compliance with the requirements for brokering;
- provide the details of the systems which it has in place to ensure such compliance.

An annual compliance report will need to be submitted in relation to any application made before 31 March 2014, the date of the application and in relation to each subsequent reporting year, by 30 April following the end of that year. The annual compliance report will be subject to a variation procedure so that the broker can change the original details provided.

The broker must without delay notify the licensing authority of any changes to the matters in respect of which evidence has been supplied in relation to the compliance report which might affect compliance with the requirements of brokering.

The broker must immediately inform MHRA and the marketing authorisation holder, of medicinal products they are offered which they identify as falsified or suspect to be falsified.

Good Distribution Practice

The Commission's guidelines on good distribution practice, referred to in Article 84 of Directive 2001/83/EC have been updated to include specific provisions for brokering, see Chapter 8.

Management of Recall Activity by Brokers

Chapter 10 of the EU GDP Guidelines deals with brokers, and includes the requirement for the broker to have in place "...an emergency plan which ensures effective recall of medicinal products from the market...". The guidelines go on to specify that brokers should have a written procedure for supporting recalls, alongside an obligation to maintain relevant documentation.

As brokers do not hold or handle stock the scope of their recall activities are somewhat more limited than wholesalers.

A broker should:

- follow the detail of the recall message;
- identify those persons or companies involved in relevant brokered deals;
- directly contact those customers, making them aware of the details of the recall;
- keep recall notices open for an appropriate period, so as to capture any affected stock still moving through the supply chain.

As with wholesalers all recall activities should be documented at the time they occur and at the conclusion of all recall activities the Broker should produce a report, making an objective assessment of whether the recall process achieved its objectives and identifying any areas requiring improvement.

Supply and brokering to countries outside of the UK

Where wholesale stock has been supplied or brokered to countries outside of the UK, there are the same obligations to contact the relevant customers to make them aware of the recall. It should be noted that certainly within the EEA, dependent on the nature of the recall and its impact on product availability and patient health, different countries may take varying levels of action. The wholesaler or broker should confirm with the relevant competent authority the particular terms of a recall in a particular territory.

Manufacture, Importation and Distribution of Active Substances

CHAPTER

10

EU Legislation on Manufacture, Importation and Distribution of Active Substances

Contents

DIRECTIVE 2001/83/EC, TITLE IV

Directive 2001/83/EC of the European Parliament and of the Council of 6 November 2001 on the Community code relating to medicinal products for human use

Editor's note Title IV of this directive is reproduced below. Reference should be made to the full Directive for the preamble, definitions and the general and final provisions.

Please note: only European Union legislation printed in the paper edition of the Official Journal of the European Union is deemed authentic.

Title IV

Article 46a

1 For the purposes of this Directive, manufacture of active substances used as starting materials shall include both total and partial manufacture or import of an active substance used as a starting material as defined in Part I, point 3.2.1.1 (b) Annex I, and the various processes of dividing up, packaging or presentation prior to its incorporation into a medicinal product, including repackaging or relabelling, such as are carried out by a distributor of starting materials.

2 The Commission shall be empowered to adapt paragraph 1 to take account of scientific and technical progress. That measure, designed to amend non-essential elements of this Directive, shall be adopted in accordance with the regulatory procedure with scrutiny referred to in Article 121(2a).

Article 46b

1 Member States shall take appropriate measures to ensure that the manufacture, import and distribution on their territory of active substances, including active substances that are intended for export, comply with good manufacturing practice and good distribution practices for active substances.

2 Active substances shall only be imported if the following conditions are fulfilled:

(a) the active substances have been manufactured in accordance with standards of good manufacturing practice at least equivalent to those laid down by the Union pursuant to the third paragraph of Article 47; and

(b) the active substances are accompanied by a written confirmation from the competent authority of the exporting third country of the following:

(i) the standards of good manufacturing practice applicable to the plant manufacturing the exported active substance are at least equivalent to those laid down by the Union pursuant to the third paragraph of Article 47;

(ii) the manufacturing plant concerned is subject to regular, strict and transparent controls and to the effective enforcement of good manufacturing practice, including repeated and unannounced inspections, so as to ensure a protection of public health at least equivalent to that in the Union; and

(iii) in the event of findings relating to non-compliance, information on such findings is supplied by the exporting third country to the Union without any delay.

This written confirmation shall be without prejudice to the obligations set out in Article 8 and in point (f) of Article 46.

3 The requirement set out in point (b) of paragraph 2 of this Article shall not apply if the exporting country is included in the list referred to in Article 111b.

4 Exceptionally and where necessary to ensure the availability of medicinal products, when a plant manufacturing an active substance for export has been inspected by a Member State and was found to comply with the principles and guidelines of good manufacturing practice laid down pursuant to the third paragraph of Article 47, the requirement set out in point (b) of paragraph 2 of this Article may be waived by any Member State for a period not exceeding the validity of the certificate of Good Manufacturing Practice. Member States that make use of the possibility of such waiver, shall communicate this to the Commission.

Article 47

The principles and guidelines of good manufacturing practices for medicinal products referred to in Article 46(f) shall be adopted in the form of a directive. That measure, designed to amend non-essential elements of this Directive by supplementing it, shall be adopted in accordance with the regulatory procedure with scrutiny referred to in Article 121(2a).

Detailed guidelines in line with those principles will be published by the Commission and revised necessary to take account of technical and scientific progress.

The Commission shall adopt, by means of delegated acts in accordance with Article 121a and subject to the conditions laid down in Articles 121b and 121c, the principles and guidelines of good manufacturing practice for active substances referred to in the first paragraph of point (f) of Article 46 and in Article 46b.

The principles of good distribution practices for active substances referred to in the first paragraph of point (f) of Article 46 shall be adopted by the Commission in the form of guidelines.

The Commission shall adopt guidelines on the formalised risk assessment for ascertaining the appropriate good manufacturing practice for excipients referred to in the second paragraph of point (f) of Article 46.

Article 52a

1 Importers, manufacturers and distributors of active substances who are established in the Union shall register their activity with the competent authority of the Member State in which they are established.

2 The registration form shall include, at least, the following information:

(i) name or corporate name and permanent address;
(ii) the active substances which are to be imported, manufactured or distributed;
(iii) particulars regarding the premises and the technical equipment for their activity.

3 The persons referred to in paragraph 1 shall submit the registration form to the competent authority at least 60 days prior to the intended commencement of their activity.

4 The competent authority may, based on a risk assessment, decide to carry out an inspection. If the competent authority notifies the applicant within 60 days of the receipt of the registration form that an inspection will be carried out, the activity shall not begin before the competent authority has notified the applicant that he may commence the activity. If within 60 days of the receipt of the registration form the competent authority has not notified the applicant that an inspection will be carried out, the applicant may commence the activity.

5 The persons referred to in paragraph 1 shall communicate annually to the competent authority an inventory of the changes which have taken place as regards the information provided in the registration form. Any changes that may have an impact on the quality or safety of the active substances that are manufactured, imported or distributed must be notified immediately.

6 Persons referred to in paragraph 1 who had commenced their activity before 2 January 2013 shall submit the registration form to the competent authority by 2 March 2013.

7 Member States shall enter the information provided in accordance with paragraph 2 of this Article in the Union database referred to in Article 111(6).

8 This Article shall be without prejudice to Article 111.

Article 53

The provisions of this Title shall also apply to homeopathic medicinal products.

CHAPTER

11

UK Legislation on the Manufacture, Importation and Distribution of Active Substances

Contents

The Human Medicines Regulations 2012 (SI 2012/1916) 219
Citation and commencement 219
General interpretation 220
Interpretation 221
Criteria for importation, manufacture or distribution of active substances 221
Registration in relation to active substances 221
Requirements for registration as an importer, manufacturer or distributor of an active substance 222
Provision of information 224
Schedule 7A Information to be provided for registration as an importer, manufacturer or distributor of active substances 225

The Human Medicines Regulations 2012 (SI 2012/1916)

Editor's note These extracts from the Human Medicines Regulations 2012 [SI 2012/1916] as amended are presented for the reader's convenience. Reproduction is with the permission of HMSO and the Queen's Printer for Scotland. For any definitive information reference must be made to the original Regulations. The numbering and content within this section correspond with the regulations set out in the published Statutory Instrument [SI 2012/1916] as amended.

Citation and commencement

1 (1) These Regulations may be cited as the Human Medicines Regulations 2012.

(2) These Regulations come into force on 14th August 2012.

General interpretation

8 (1) In these Regulations (unless the context otherwise requires)- "active substance" means any substance or mixture of substances intended to be used in the manufacture of a medicinal product and that, when used in its production, becomes an active ingredient of that product intended to exert a pharmacological, immunological or metabolic action with a view to restoring, correcting or modifying physiological functions or to make a medical diagnosis;

"assemble", in relation to a medicinal product or an active substance, includes the various processes of dividing up, packaging and presentation of the product or substance, and "assembly" has a corresponding meaning;

"excipient" means any constituent of a medicinal product other than the active substance and the packaging material;

"export" means export, or attempt to export, from the United Kingdom, whether by land, sea or air;

"falsified medicinal product" means any medicinal product with a false representation of:

(a) its identity, including its packaging and labelling, its name or its composition (other than any unintentional quality defect) as regards any of its ingredients including excipients and the strength of those ingredients;

(b) its source, including its manufacturer, its country of manufacturing, its country of origin or its marketing authorisation holder; or

(c) its history, including the records and documents relating to the distribution channels used;

"import" means import, or attempt to import, into the United Kingdom, whether by land, sea or air;

(8) References in these Regulations to:

(a) good manufacturing practice for active substances relate to the principles and guidelines for good manufacturing practice adopted by the European Commission under the third paragraph of Article 47[1] of the 2001 Directive;

(b) good distribution practice for active substances relate to the guidelines on good distribution practices for active substances adopted by the European Commission under the fourth paragraph of Article 47 of the 2001 Directive.

[1] Paragraphs 3 and 4 of Article 47 were substituted by Directive 2011/62/EU of the European Parliament and of the Council (OJ No L 174, 1.7.2011, p74).

Interpretation

A17. In this Part "manufacture", in relation to an active substance, includes any process carried out in the course of making the substance and the various processes of dividing up, packaging, and presentation of the active substance.

Criteria for importation, manufacture or distribution of active substances

45M. (1) A person may not:
(a) import;
(b) manufacture; or
(c) distribute,
an active substance unless that person is registered with the licensing authority in accordance with regulation 45N and the requirements in regulation 45O are met.
(2) Paragraph (1) applies in relation to an active substance which is to be used in an investigational medicinal product only:
(a) if the product has a marketing authorisation, Article 126a authorisation, certificate of registration or traditional herbal registration; and
(b) to the extent that the manufacture of the active substance is in accordance with the terms and conditions of that authorisation, certificate or registration.
(3) Paragraph (1)(a) does not apply to a person who, in connection with the importation of an active substance from a state other than an EEA state:
(a) provides facilities solely for transporting the active substance; or
(b) acting as an import agent, imports the active substance solely to the order of another person who holds a certificate of good manufacturing practice issued by the licensing authority.

Registration in relation to active substances

45N. (1) For registration in relation to active substances, the licensing authority must have received a valid registration form from the applicant for import, manufacture or, as the case may be, distribution of the active substance and:
(a) 60 days have elapsed since receipt and the licensing authority have not notified the applicant that an inspection will be carried out; or
(b) the licensing authority:

(i) notified the applicant within 60 days of receipt of a registration form that an inspection will be carried out; and
(ii) within 90 days of that inspection the licensing authority have issued that person with a certificate of good manufacturing practice or, as the case may be, of good distribution practice; and

(c) that person has not instructed the licensing authority to end that person's registration.

(2) The person applying for registration under paragraph (1) must notify the licensing authority of any changes which have taken place as regards the information in the registration form:
(a) immediately where such changes may have an impact on quality or safety of the active substances that are manufactured, imported or distributed;
(b) in any other case, on each anniversary of the receipt of the application form by the licensing authority.

(3) For the purpose of paragraph (2), changes which are notified in accordance with that paragraph shall be treated as incorporated in the application form.

(4) Any notification to the licensing authority under paragraph (2) must be accompanied by the appropriate fee in accordance with the Fees Regulations.

(5) A registration form is valid for the purpose of paragraph (1) if:
(a) it is provided to the licensing authority; and
(b) is completed in the way and form specified in Schedule 7A.

(6) Paragraph (1) does not apply until 20th October 2013 in relation to a person who had, before 20th August 2013, commenced the activity for which the person would, apart from this provision, need to send a registration form to the licensing authority.

Requirements for registration as an importer, manufacturer or distributor of an active substance

450. (1) Where the Commission has adopted principles and guidelines of good manufacturing practice under the third paragraph of Article 47[2] of the 2001 Directive which applies to an active substance manufactured in the UK, the manufacturer must comply with good manufacturing practice in relation to that active substance.

(2) Where the Commission has adopted principles and guidelines of good distribution practice under the fourth paragraph of Article 47 of the

[2] Article 47 was amended by Directive 2011/62/EU of the European Parliament and of the Council (OJ No L 174, 1.7.2011, p74).

2001 Directive which applies to an active substance distributed in the United Kingdom, the distributor must comply with good distribution practice in relation to that active substance.

(3) Without prejudice to regulation 37(4) (manufacture and assembly in relation to active substances) and paragraph 9A of Schedule 8 (material to accompany an application for a UK marketing authorisation in relation to an active substance), where the Commission has adopted principles and guidelines of good manufacturing practice under the third paragraph of Article 47 of the 2001 Directive which applies to an active substance imported into the UK and where an active substance is imported from a third country:
 (a) the importer must comply with good manufacturing practice and good distribution practice in relation to the active substance;
 (b) the active substances must have been manufactured in accordance with standards which are at least equivalent to good manufacturing practice; and
 (c) the active substances must be accompanied by a written confirmation from the competent authority of the exporting third country of the following:
  (i) the standards of manufacturing practice applicable to the plant manufacturing the exported active substance are at least equivalent to good manufacturing practice,
  (ii) the manufacturing plant concerned is subject to regular, strict and transparent controls and to the effective enforcement of standards of manufacturing practice at least equivalent to good manufacturing practice, including repeated and unannounced inspections, so as to ensure a protection of public health at least equivalent to that in the Union, and
  (iii) in the event of findings relating to non-compliance, information on such findings is supplied by the exporting third country to the Union without any delay.

(4) Paragraph (3)(c) does not apply:
 (a) where the country from where the active substance is exported is included in the list referred to in Article 111b of the 2001 Directive; or
 (b) for a period not exceeding the validity of the certificate of good manufacturing practice, where:
  (i) in relation to a plant where active substances are manufactured where the competent authority of a member State has found, upon inspection, that a plant complies with the principles and guidelines of good manufacturing practice, and
  (ii) the licensing authority is of the opinion that it is necessary to waive the requirement to ensure availability of the active substance.

(5) The criteria in this regulation apply regardless of whether an active substance is intended for export.

Provision of information

45P. (1) In this regulation:
"R" means a person who is, or has applied to the licensing authority to become, a registered importer, manufacturer or distributor of active substances;
"reporting year" means a period of twelve months ending on 31st March.

(2) On or before the date specified in paragraph (3), R must submit a report to the licensing authority which:
(a) includes a declaration that R has in place an appropriate system to ensure compliance with regulations 45N, 45O and this regulation; and
(b) details the system which R has in place to ensure such compliance.

(3) The date specified for the purposes of this paragraph is:
(a) in relation to any application made before 31st March 2014, the date of the application; and
(b) in relation to each subsequent reporting year, 30th April following the end of that year.

(4) R must without delay notify the licensing authority of any changes to the matters in respect of which evidence has been supplied in relation to paragraph (2) which might affect compliance with the requirements of this Chapter.

(5) Any report or notification to the licensing authority under paragraph (2) or (4) must be accompanied by the appropriate fee in accordance with the Fees Regulations.

(6) The licensing authority may give a notice to R, requiring R to provide information of a kind specified in the notice within the period specified in the notice.

(7) A notice under paragraph (6) may not be given to R unless it appears to the licensing authority that it is necessary for the licensing authority to consider whether the registration should be varied, suspended or removed from the active substance register.

(8) A notice under paragraph (6) may specify information which the licensing authority thinks necessary for considering whether the registration should be varied, suspended or removed from the active substance register.

Schedule 7A – Information to be provided for registration as an importer, manufacturer or distributor of active substances

(1) The name and address of the applicant.
(2) The name and address of the person (if any) making the application on the applicant's behalf.
(3) The address of each of the premises where any operations to which the registration relates are to be carried out.
(4) The address of any premises not mentioned by virtue of the above requirement, where:
  (a) the applicant proposes to keep any living animals, from which substance(s) used in the production of the active substance(s) to which the application relates are to be derived;
  (b) materials of animal origin from which an active substance is to be derived, as mentioned in the above sub-paragraph, are to be kept.
(5) The address of each of the premises where active substances are to be stored, or from which active substances are to be distributed.
(6) The address of each of the premises where any testing associated with the manufacture or assembly of active substances to which the registration relates.
(7) The name, address, qualifications and experience of the person whose duty it will be to supervise any manufacturing operations, and the name and job title of the person to whom they report.
(8) The name, address, qualifications and experience of the person who will have responsibility for the quality control of active substances, and the name and job title of the person to whom they report.
(9) The name, address, qualifications and experience of the person whose duty it will be to supervise any importation, storage or distribution operations, and the name and job title of the person to whom they report.
(10) The name, address and qualifications of the person to be responsible for any animals kept as mentioned in paragraph 4(a).
(11) The name, address and qualifications of the person to be responsible for the culture of any living tissue for use in the manufacture of an active substance.
(12) For each active substance to be manufactured, imported, or distributed:
  (a) the CAS registration number[3] assigned to that active substance by the Chemical Abstracts Service, a division of the American Chemical Society;

[3] Further information is available from the website of the Chemical Abstracts Service at www.cas.org.

(b) where applicable, the Anatomical Therapeutic Category code[4] assigned to that active substance under the Anatomical Therapeutic Chemical Classification System used for the classification of drugs by the World Health Organization's Collaborating Centre for Drug Statistics Methodology;
(c) either:
(i) the International Union of Pure and Applied Chemistry nomenclature, or
(ii) the common name; and
(d) the intended quantities of each active substance to be manufactured, imported or distributed.

(13) Details of the operations to which the registration relates, including a statement of whether they include:
(a) the manufacture of active substances;
(b) the importation of active substances from third countries;
(c) the storage of active substances; or
(d) the distribution of active substances.

(14) A statement of the facilities and equipment available at each of the premises where active substances are to be manufactured, stored or distributed.

(15) A statement as to whether the particular active substances are intended for:
(a) use in a medicinal product with an EU marketing authorisation;
(b) use in a special medicinal product; or
(c) export to a third country.

(16) A separate statement in respect of each of the premises mentioned in the application of:
(a) the manufacturing, storage or distribution operations carried out at those sites, and the specific active substances to which those activities relate; and
(b) the equipment available at those premises for carrying out those activities.

(17) A statement of the authority conferred on the person responsible for quality control to reject unsatisfactory active substances.

(18) A description of the arrangements for the identification and storage of materials before and during the manufacture of active substances.

(19) A description of the arrangements for the identification and storage of active substances.

(20) A description of the arrangements at each of the premises where the applicant proposes to store active substances for ensuring, as far as practicable, the turn-over of stocks of active substances.

[4] Further information is available from the website of the WHO Collaborating Centre for Drug Statistics Methodology at www.whocc.no.

(21) A description of the arrangements for maintaining:
  (a) production records, including records of manufacture and assembly;
  (b) records of analytical and other tests used in the course of manufacture or assembly for ensuring compliance of materials used in manufacture, or of active substances, with the specification for such materials or active substances;
  (c) records of importation;
  (d) records of storage and distribution.

(22) A description of the arrangements for keeping reference samples of:
  (a) materials used in the manufacture of active substances; and
  (b) active substances.

(23) Where the application relates to active substances intended for use in an advanced therapy medicinal product, an outline of the arrangements for maintaining records to allow traceability containing sufficient detail to enable the linking of an active substance to the advanced therapy medicinal product it was used in the manufacture of and vice versa.

(24) Details of:
  (a) any manufacturing, importation, storage or distribution operations, other than those to which the application for registration relates, carried on by the applicant on or near each of the premises, and
  (b) the substances or articles to which those operations relate.

CHAPTER

12

Guidelines of 19 March 2015 on Principles of Good Distribution Practice of Active Substances for Medicinal Products for Human Use (2015/C 95/01)

Contents

Introduction 228
Chapter 1 – Scope 229
Chapter 2 – Quality System 229
Chapter 3 – Personnel 230
Chapter 4 – Documentation 230
Procedures 231
Records 231
Chapter 5 – Premises and Equipment 232
Chapter 6 – Operations 232
Orders 232
Receipt 232
Storage 233
Deliveries to customers 234
Transfer of information 234
Chapter 7 – Returns, Complaints and Recalls 235
Returns 235
Complaints and recalls 236
Chapter 8 – Self-inspections 237
Annex 238

Introduction

These guidelines are based on the fourth paragraph of Article 47 of Directive 2001/83/EC[1].

They follow the same principles that underlie the guidelines of EudraLex Volume 4, Part II, Chapter 17, with regard to the distribution of active substances and the Guidelines of 5 November 2013 on Good Distribution Practice of medicinal products for human use[2].

These guidelines provide stand-alone guidance on Good Distribution Practice (GDP) for importers and distributors of active substances for medicinal products for human use. They complement the rules on

[1] Directive 2001/83/EC of the European Parliament and of the Council of 6 November 2001 on the Community code relating to medicinal products for human use (OJ L 311, 28.11.2001, p. 67).

[2] OJ C 343, 23.11.2013, p. 1.

distribution set out in the guidelines of EudraLex Volume 4, Part II, and apply also to distributors of active substances manufactured by themselves.

Any manufacturing activities in relation to active substances, including re-packaging, re-labelling or dividing up, are subject to Commission Delegated Regulation (EU) No 1252/2014[3] and EudraLex Volume 4, Part II.

Additional requirements apply to the importation of active substances, as laid down in Article 46b of Directive 2001/83/EC.

Distributors of active substances for medicinal products for human use should follow these guidelines as of 21 September 2015.

CHAPTER 1 – SCOPE

1.1 These guidelines apply to distribution of active substances, as defined in Article 1(3a) of Directive 2001/83/EC, for medicinal products for human use. According to that provision, an active substance is any substance or mixture of substances intended to be used in the manufacture of a medicinal product and that, when used in its production, becomes an active ingredient of that product intended to exert a pharmacological, immunological or metabolic action with a view to restoring, correcting or modifying physiological functions or to make a medical diagnosis.

1.2 For the purpose of these guidelines, distribution of active substances shall comprise all activities consisting of procuring, importing, holding, supplying or exporting active substances, apart from brokering.

1.3 These guidelines do not apply to intermediates of active substances.

CHAPTER 2 – QUALITY SYSTEM

2.1 Distributors of active substances should develop and maintain a quality system setting out responsibilities, processes and risk management principles. Examples of the processes and applications of quality risk management can be found in EudraLex Volume 4, Part III: GMP related documents, ICH guideline Q9 on Quality Risk Management (ICH Q9).

2.2 The quality system should be adequately resourced with competent personnel, and suitable and sufficient premises, equipment and facilities. It should ensure that:

[3] Commission Delegated Regulation (EU) No 1252/2014 of 28 May 2014 supplementing Directive 2001/83/EC of the European Parliament and of the Council with regard to principles and guidelines of good manufacturing practice for active substances for medicinal products for human use (OJ L 337, 25.11. 2014, p. 1).

(i) active substances are procured, imported, held, supplied or exported in a way that is compliant with the requirements of GDP for active substances;
(ii) management responsibilities are clearly specified;
(iii) active substances are delivered to the right recipients within a satisfactory time period;
(iv) records are made contemporaneously;
(v) deviations from established procedures are documented and investigated;
(vi) appropriate corrective and preventive actions, commonly known as 'CAPA', are taken to correct deviations and prevent them in line with the principles of quality risk management;
(vii) changes that may affect the storage and distribution of active substances are evaluated.

2.3 The size, structure and complexity of the distributor's activities should be taken into consideration when developing or modifying the quality system.

CHAPTER 3 – PERSONNEL

3.1 The distributor should designate a person at each location where distribution activities are performed who should have defined authority and responsibility for ensuring that a quality system is implemented and maintained. The designated person should fulfil his responsibilities personally. The designated person can delegate duties but not responsibilities.

3.2 The responsibilities of all personnel involved in the distribution of active substances should be specified in writing. The personnel should be trained on the requirements of GDP for active substances. They should have the appropriate competence and experience to ensure that active substances are properly handled, stored and distributed.

3.3 Personnel should receive initial and continuing training relevant to their role, based on written procedures and in accordance with a written training programme.

3.4 A record of all training should be kept, and the effectiveness of training should be periodically assessed and documented.

CHAPTER 4 – DOCUMENTATION

4.1 Documentation comprises all written procedures, instructions, contracts, records and data, in paper or in electronic form. Documentation should be readily available or retrievable. All documentation related to compliance of the distributor with these guidelines should be made available on request of competent authorities.

4.2 Documentation should be sufficiently comprehensive with respect to the scope of the distributor's activities and in a language understood by personnel. It should be written in clear, unambiguous language and be free from errors.

4.3 Any alteration made in the documentation should be signed and dated; the alteration should permit the reading of the original information. Where appropriate, the reason for the alteration should be recorded.

4.4 Each employee should have ready access to all necessary documentation for the tasks executed.

Procedures

4.5 Written procedures should describe the distribution activities which affect the quality of the active substances. This could include receipt and checking of deliveries, storage, cleaning and maintenance of the premises (including pest control), recording of the storage conditions, security of stocks on site and of consignments in transit, withdrawal from saleable stock, handling of returned products, recall plans, etc.

4.6 Procedures should be approved, signed and dated by the person responsible for the quality system.

4.7 Attention should be paid to the use of valid and approved procedures. Documents should be reviewed regularly and kept up to date. Version control should be applied to procedures. After revision of a document a system should exist to prevent inadvertent use of the superseded version. Superseded or obsolete procedures should be removed from workstations and archived.

Records

4.8 Records should be clear, be made at the time each operation is performed and in such a way that all significant activities or events are traceable. Records should be retained for at least 1 year after the expiry date of the active substance batch to which they relate. For active substances with retest dates, records should be retained for at least 3 years after the batch is completely distributed.

4.9 Records should be kept of each purchase and sale, showing the date of purchase or supply, name of the active substance, batch number and quantity received or supplied, and name and address of the supplier and of the original manufacturer, if not the same, or of the shipping agent and/or the consignee. Records should ensure the traceability of the origin and destination of products, so that all the suppliers of, or those supplied with, an active substance can be identified. Records that should be retained and be available include:

(i) identity of supplier, original manufacturer, shipping agent and/or consignee;
(ii) address of supplier, original manufacturer, shipping agent and/or consignee;
(iii) purchase orders;
(iv) bills of lading, transportation and distribution records;
(v) receipt documents;
(vi) name or designation of active substance;
(vii) manufacturer's batch number;
(viii) certificates of analysis, including those of the original manufacturer;
(ix) retest or expiry date.

CHAPTER 5 – PREMISES AND EQUIPMENT

5.1 Premises and equipment should be suitable and adequate to ensure proper storage, protection from contamination, e.g. narcotics, highly sensitising materials, materials of high pharmacological activity or toxicity, and distribution of active substances. They should be suitably secure to prevent unauthorised access. Monitoring devices that are necessary to guarantee the quality attributes of the active substance should be calibrated according to an approved schedule against certified traceable standards.

CHAPTER 6 – OPERATIONS

Orders

6.1 Where active substances are procured from a manufacturer, importer or distributor established in the EU, that manufacturer, importer or distributor should be registered according to Article 52a of Directive 2001/83/EC.

Receipt

6.2 Areas for receiving active substances should protect deliveries from prevailing weather conditions during unloading. The reception area should be separate from the storage area. Deliveries should be examined at receipt in order to check that:

(i) containers are not damaged;
(ii) all security seals are present with no sign of tampering;
(iii) correct labelling, including correlation between the name used by the supplier and the in-house name, if these are different;
(iv) necessary information, such as a certificate of analysis, is available; and
(v) the active substance and the consignment correspond to the order.

6.3 Active substances with broken seals, damaged packaging, or suspected of possible contamination should be quarantined either physically or using an equivalent electronic system and the cause of the issue investigated.

6.4 Active substances subject to specific storage measures, e.g. narcotics and products requiring a specific storage temperature or humidity, should be immediately identified and stored in accordance with written instructions and with relevant legislative provisions.

6.5 Where the distributor suspects that an active substance procured or imported by him is falsified, he should segregate it either physically or using an equivalent electronic system and inform the national competent authority of the country in which he is registered.

6.6 Rejected materials should be identified and controlled and quarantined to prevent their unauthorised use in manufacturing and their further distribution. Records of destruction activities should be readily available.

Storage

6.7 Active substances should be stored under the conditions specified by the manufacturer, e.g. controlled temperature and humidity when necessary, and in such a manner to prevent contamination and/or mix up. The storage conditions should be monitored and records maintained. The records should be reviewed regularly by the person responsible for the quality system.

6.8 When specific storage conditions are required, the storage area should be qualified and operated within the specified limits.

6.9 The storage facilities should be clean and free from litter, dust and pests. Adequate precautions should be taken against spillage or breakage, attack by micro-organisms and cross-contamination.

6.10 There should be a system to ensure stock rotation, e.g. 'first expiry (retest date), first out', with regular and frequent checks that the system is operating correctly. Electronic warehouse management systems should be validated.

6.11 Active substances beyond their expiry date should be separated, either physically or using an equivalent electronic system, from approved stock and not be supplied.

6.12 Where storage or transportation of active substances is contracted out, the distributor should ensure that the contract acceptor knows and follows the appropriate storage and transport conditions. There must be a written contract between the contract giver and contract acceptor, which clearly establishes the duties of each party. The contract acceptor should not subcontract any of the work entrusted to him under the contract without the contract giver's written authorisation.

Deliveries to customers

6.13 Supplies within the EU should be made only by distributors of active substances registered according to Article 52a of Directive 2001/83/EC to other distributors, manufacturers or to dispensing pharmacies.

6.14 Active substances should be transported in accordance with the conditions specified by the manufacturer and in a manner that does not adversely affect their quality. Product, batch and container identity should be maintained at all times. All original container labels should remain readable.

6.15 A system should be in place by which the distribution of each batch of active substance can be readily identified to permit its recall.

Transfer of information

6.16 Any information or event that the distributor becomes aware of, which have the potential to cause an interruption to supply, should be notified to relevant customers.

6.17 Distributors should transfer all product quality or regulatory information received from an active substance manufacturer to the customer and from the customer to the active substance manufacturer.

6.18 The distributor who supplies the active substance to the customer should provide the name and address of the original active substance manufacturer and the batch number(s) supplied. A copy of the original certificate of analysis from the manufacturer should be provided to the customer.

6.19 The distributor should also provide the identity of the original active substance manufacturer to competent authorities upon request. The original manufacturer can respond to the competent authority directly or through its authorised agents. (In this context 'authorised' refers to authorised by the manufacturer.)

6.20 The specific guidance for certificates of analysis is detailed in Section 11.4 of Part II of Eudralex Volume 4.

CHAPTER 7 – RETURNS, COMPLAINTS AND RECALLS

Returns

7.1 Returned active substances should be identified as such and quarantined pending investigation.

7.2 Active substances which have left the care of the distributor, should only be returned to approved stock if all of the following conditions are met:

(i) the active substance is in the original unopened container(s) with all original security seals present and is in good condition;
(ii) it is demonstrated that the active substance has been stored and handled under proper conditions. Written information provided by the customer should be available for this purpose;
(iii) the remaining shelf life period is acceptable;
(iv) the active substance has been examined and assessed by a person trained and authorised to do so;
(v) no loss of information/traceability has occurred.

This assessment should take into account the nature of the active substance, any special storage conditions it requires, and the time elapsed since it was supplied. As necessary and if there is any doubt about the quality of the returned active substance, advice should be sought from the manufacturer.

7.3 Records of returned active substances should be maintained. For each return, documentation should include:

(i) name and address of the consignee returning the active substances;
(ii) name or designation of active substance, active substance batch number and quantity returned;
(iii) reason for return;
(iv) use or disposal of the returned active substance and records of the assessment performed.

7.4 Only appropriately trained and authorised personnel should release active substances for return to stock. Active substances returned to saleable stock should be placed such that the stock rotation system operates effectively.

Complaints and recalls

7.5 All complaints, whether received orally or in writing, should be recorded and investigated according to a written procedure. In the event of a complaint about the quality of an active substance the distributor should review the complaint with the original active substance manufacturer in order to determine whether any further action, either with other customers who may have received this active substance or with the competent authority, or both, should be initiated. The investigation into the cause for the complaint should be conducted and documented by the appropriate party.

7.6 Complaint records should include:

(i) name and address of complainant;
(ii) name, title, where appropriate, and phone number of person submitting the complaint;
(iii) complaint nature, including name and batch number of the active substance;
(iv) date the complaint is received;
(v) action initially taken, including dates and identity of person taking the action;
(vi) any follow-up action taken;
(vii) response provided to the originator of complaint, including date response sent;
(viii) final decision on active substance batch.

7.7 Records of complaints should be retained in order to evaluate trends, product related frequencies, and severity with a view to taking additional, and if appropriate, immediate corrective action. These should be made available during inspections by competent authorities.

7.8 Where a complaint is referred to the original active substance manufacturer, the record maintained by the distributor should include any response received from the original active substance manufacturer, including date and information provided.

7.9 In the event of a serious or potentially life-threatening situation, local, national, and/or international authorities should be informed and their advice sought.

7.10 There should be a written procedure that defines the circumstances under which a recall of an active substance should be considered.

7.11 The recall procedure should designate who should be involved in evaluating the information, how a recall should be initiated, who should be informed about the recall, and how the recalled material should be treated. The designated person (cf. Section 3.1) should be involved in recalls.

CHAPTER 8 – SELF-INSPECTIONS

8.1 The distributor should conduct and record self-inspections in order to monitor the implementation of and compliance with these guidelines. Regular self-inspections should be performed in accordance with an approved schedule.

Annex

Glossary of terms applicable to these guidelines

Terms	Definition
Batch	A specific quantity of material produced in a process or series of processes so that it is expected to be homogeneous within specified limits. In the case of continuous production, a batch may correspond to a defined fraction of the production. The batch size can be defined either by a fixed quantity or by the amount produced in a fixed time interval.
Batch number	A unique combination of numbers, letters and/or symbols that identifies a batch (or lot) and from which the production and distribution history can be determined.
Brokering of active substances	All activities in relation to the sale or purchase of active substances that do not include physical handling and that consist of negotiating independently and on behalf of another legal or natural person.
Calibration	The demonstration that a particular instrument or device produces results within specified limits by comparison with those produced by a reference or traceable standard over an appropriate range of measurements.
Consignee	The person to whom the shipment is to be delivered whether by land, sea or air.
Contamination	The undesired introduction of impurities of a chemical or microbiological nature, or of foreign matter, into or onto a raw material, intermediate, or active substance during production, sampling, packaging or repackaging, storage or transport.
Distribution of active substances	All activities consisting of procuring, importing, holding, supplying or exporting of active substances, apart from brokering.
Deviation	Departure from an approved instruction or established standard.
Expiry date	The date placed on the container/labels of an active substance designating the time during which the active substance is expected to remain within established shelf life specifications if stored under defined conditions, and after which it should not be used.

Terms	Definition
Falsified active substance	Any active substance with a false representation of: **(a)** its identity, including its packaging and labelling, its name or its components as regards any of the ingredients and the strength of those ingredients; **(b)** its source, including its manufacturer, its country of manufacture, its country of origin; or **(c)** its history, including the records and documents relating to the distribution channels used.
Holding	Storing active substances.
Procedure	A documented description of the operations to be performed, the precautions to be taken and measures to be applied directly or indirectly related to the distribution of an active substance.
Procuring	Obtaining, acquiring, purchasing or buying active substances from manufacturers, importers or other distributors.
Quality risk management	A systematic process for the assessment, control, communication and review of risks to the quality of an active substance across the product lifecycle.
Quality system	The sum of all aspects of a system that implements quality policy and ensures that quality objectives are met (ICH Q9).
Quarantine	The status of materials isolated physically or by other effective means pending a decision on the subsequent approval or rejection.
Retest date	The date when a material should be re-examined to ensure that it is still suitable for use.
Supplying	All activities of providing, selling, donating active substances to distributors, pharmacists, or manufacturers of medicinal products.
Signed (signature)	The record of the individual who performed a particular action or review. This record can be initials, full handwritten signature, personal seal, or authenticated and secure electronic signature.
Transport (transportation)	Moving active substances between two locations without storing them for unjustified periods of time.
Validation	A documented program that provides a high degree of assurance that a specific process, method, or system will consistently produce a result meeting pre-determined acceptance criteria.

CHAPTER

13

UK Guidance on the Manufacture, Importation and Distribution of Active Substances

Contents

Introduction 240
Registration 241
Conditions of Registration as a Manufacturer, Importer or Distributor of an Active Substance 244
GMP for Active Substances 245
GDP for Active Substances 246
Written Confirmation 246
Template for the 'written confirmation' for active substances exported to the European Union for medicinal products for human use, in accordance with Article 46b(2)(b) of Directive 2001/83/EC 247
Annex 248
Waiver from Written Confirmation 249
Procedure for Active Substance Importation 250
Procedure for Waiver from Written Confirmation 252
National Contingency Guidance 253
National Contingency Guidance Submission Template 255
PART A: Finished Product Manufacturer to which this declaration applies 255
PART B: Concerned Third Country Active Substance Manufacturing Site 255
PART C: Basis of declaration in lieu of full compliance with Article 46b(2) 256
PART D: Declaration 257
PART E: Name and Signature of QP responsible for this Declaration 258

Introduction

EU Directive 2001/83/EC lays down the rules for the manufacture, import, marketing and supply of medicinal products and ensures the functioning of the internal market for medicinal products while safeguarding a high level of protection of public health in the EU.

The falsification of medicinal products is a global problem, requiring effective and enhanced international coordination and cooperation in order to ensure that anti-falsification strategies are more effective, in particular as regards sale of such products via the Internet. To that end, the Commission and Member States are cooperating closely and supporting ongoing work in international fora on this subject, such as the Council of Europe, Europol and the United Nations. In addition, the Commission, working closely with Member States, is cooperating with the competent authorities of third countries with a view to effectively combating the trade in falsified medicinal products at a global level.

Active substances are those substances which give a medicinal product its therapeutic effect. They are the Active Pharmaceutical Ingredient (API).

Falsified active substances and active substances that do not comply with applicable requirements of Directive 2001/83/EC pose serious risks to public health.

The Falsified Medicines Directive 2011/62/EU amends Directive 2001/83/EC in order to facilitate the enforcement of and control of compliance with Union rules relating to active substances. It makes a number of significant changes to the controls on active substances intended for use in the manufacture of a medicinal product for human use.

A number of new terms have been introduced into the 2001 Directive by the Falsified Medicines Directive, including "falsified medicinal product" and "active substance". The aim of this is to ensure that other amendments introduced by the Falsified Medicines Directive are consistently interpreted and applied across the EU.

The 2001 Directive has been amended to permit the European Commission to adopt the following:

- the principles and guidelines of good manufacturing practice for active substances, by means of a delegated act; and
- the principles of good distribution practice for active substances, by means of adopted guidelines.

Registration

To provide a greater level of control, and transparency of supply, for active substances within the European Community manufacturers, importers and distributors of active substances have to notify the relevant competent authorities of their activities and provide certain details. In the UK this will be MHRA. The competent authority has an obligation to enter these details into a Community Database (EudraGMDP) following the determination of a successful application for registration. The competent authority may then conduct inspections against the requirements of the relevant good practices before permitting such businesses to start trading. Manufacturers, importers and distributors of active substances will not only be subject to inspection on the basis of suspicions of non-compliance, but also on the basis of risk-analysis.

Authorised manufacturers of medicinal products who also manufacture and/or import active substances, either for use in their own products or products manufactured by other companies, are not exempt from the requirement to register.

Persons who are requested to import an active substance from a non-EEA country that provide facilities solely for transporting the active substance, or where they are acting as an import agent, imports the active substance solely to the order of another person who holds a certificate of good manufacturing practice issued by the licensing authority, are not required to register.

The registration regime for manufacturers, importers and distributors of active substances will be subject to an application procedure, followed by a determination procedure completed by MHRA.

The person applying for registration must notify MHRA immediately of any changes which have taken place as regards to the information in the registration form, where such changes may have an impact on quality or safety of the active substances that are manufactured, imported or distributed. These changes shall be treated as incorporated in the application form.

MHRA must grant or refuse an application for registration within 60 working days beginning immediately after the day on which a valid application is received.

MHRA will notify the applicant within 60 days of receipt of a valid application for registration whether they intend to undertake an inspection.

The applicant may not undertake any activity before either:

- 60 days have elapsed and the applicant has not been notified of the Agency's intention to inspect, or
- following inspection the Agency has notified the applicant that they may commence their activities.

After inspection MHRA will prepare a report and communicate that report to the applicant. The applicant will have the opportunity to respond to the report. Within 90 days of an inspection MHRA shall issue an appropriate good practice certificate to the applicant, indicating that the applicant complies with the requirements of the relevant good practices. Where an applicant is found to be non-compliant with the requisite standards, a statement of non-compliance will be issued by MHRA.

If after 60 days of the receipt of the application form MHRA has not notified the applicant of their intention to carry out an inspection, the applicant may commence their business activity and regard themselves as

Registration requirements for UK companies involved in the sourcing and supply of active substances (ASs) to be used in the manufacture of human medicines.

MHRA

This flowchart may be used to determine the appropriate registration required for UK companies. Where more than one activity is carried out, the company should register for all activities that apply.

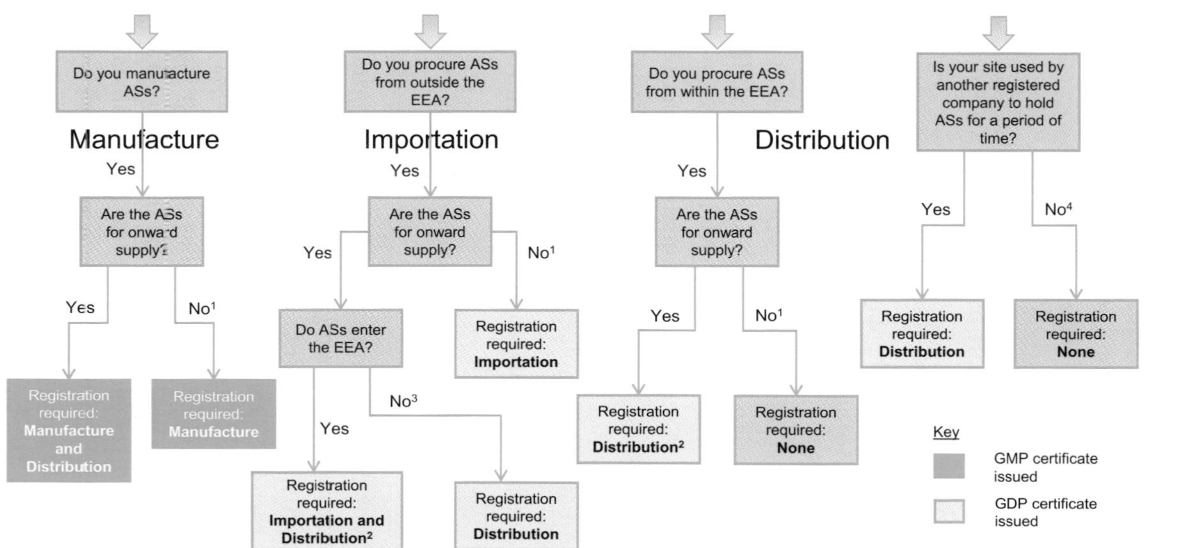

1. Site holds a manufacturer's licence and uses the ASs in the manufacture of the finished dose form.
2. Where the registration holder contracts out the physical handling of the AS, the company that physically handles the product should be named as a 3rd party site on the registration if they hold the ASs and this takes place in the UK.
   In addition, the company should hold its own registration. If the company are based in another EEA State, they should be registered with the competent authority in that State.
3. ASs procured from outside the EEA are supplied directly to customers based outside the EEA.
4. ASs are not held (stored) for any length of time and the company are simply contracted to transport the ASs.

Figure 13.1 Registration requirements for UK companies involved in the sourcing and supply of active substances (ASs) to be used in the manufacture of human medicines.

registered. MHRA will issue a certificate to the applicant and enter the details into the Community Database.

This Community Database which is publicly available will enable competent authorities in other EEA Member States or other legal entities, to establish the bona fides and compliance of manufacturers, importers and distributors of active substances established in the UK and those in other EEA territories. MHRA will investigate concerns with regards to UK registrations of non-compliance and reciprocal arrangements will apply with other EEA Member States.

Conditions of Registration as a Manufacturer, Importer or Distributor of an Active Substance

A person in the UK may not import, manufacture or distribute, an active substance for use in a licensed human medicine unless they are registered with MHRA in accordance with the Human Medicines Regulations 2012 and the respective conditions of those Regulations are met.

Registration holders must submit to MHRA an annual update of any changes to the information provided in the application. Any changes which may have an impact on the quality or safety of the active substance which the registrant is permitted to handle must be notified to the Agency immediately.

An annual compliance report will need to be submitted:

- in relation to any application made before 31 March 2013, the date of application; and
- in relation to each subsequent reporting year, 30 April following the end of that year.

Where the Commission has adopted principles and guidelines of good manufacturing practice under the third paragraph of Article 47[1] of Directive 2001/83/EC which applies to an active substance manufactured in the UK, the registered manufacturer must comply with good manufacturing practice in relation to that active substance.

Where the Commission has adopted principles and guidelines of good distribution practice under the fourth paragraph of Article 47 of Directive 2001/83/EC which applies to an active substance distributed in the UK, the registered distributor must comply with good distribution practice in relation to that active substance.

Where the Commission has adopted principles and guidelines of good manufacturing practice under the third paragraph of Article 47 of

[1] Article 47 was amended by Directive 2011/62/EU of the European Parliament and of the Council (OJ No L 174, 1.7.2011, p. 74).

Directive 2001/83/EC which applies to an active substance imported into the UK and where an active substance is imported from a third country the registered importer must comply with good distribution practice in relation to the active substance.

Under such circumstances the active substances must have been manufactured in accordance with standards which are at least equivalent to EU good manufacturing practice and when imported must be accompanied by a written confirmation from the competent authority of the exporting third country unless a waiver exists.

GMP for Active Substances

Directive 2001/83/EC has been amended to include a new definition of "active substance" which means any substance or mixture of substances intended to be used in the manufacture of a medicinal product and that, when used in its production, becomes an active ingredient of that product intended to exert a pharmacological, immunological or metabolic action with a view to restoring, correcting or modifying physiological functions or to make a medical diagnosis.

The manufacture of active substances should be subject to good manufacturing practice regardless of whether those active substances are manufactured in the Union or imported. Where active substances are manufactured in third countries it should be ensured that such substances have been manufactured to the relevant European standards of good manufacturing practice (GMP), so as to provide a level of protection of public health equivalent to that provided for by EU law.

A manufacturer or assembler of an active substance will have to comply with the principles and guidelines for GMP for active substances. Manufacture, in relation to an active substance, includes any process carried out in the course of making the substance and the various processes of dividing up, packaging, and presentation of the active substance. Assemble, in relation to an active substance, includes the various processes of dividing up, packaging and presentation of the substance, and "assembly" has a corresponding meaning. These activities will be the subject of a GMP certificate.

Importers of an active substance from a third country have to comply with the guidelines for Good Distribution Practice (GDP) in relation to the active substance. This activity will be the subject of a GDP certificate.

Distributors of an active substance within the UK which has been sourced from a manufacturer or an importer within the EU will have to comply with the guidelines for GDP for active substances. This activity will be the subject of a GDP certificate.

The 2001 Directive has been amended to permit the European Commission to adopt the following:

- the principles and guidelines of good manufacturing practice for active substances, by means of a delegated act; and
- the principles of good distribution practice for active substances, by means of adopted guidelines.

GMP for active substances is contained in Part II of the EU guidelines on Good Manufacturing Practice.

GDP for Active Substances

In March 2015 the European Commission published "Guidelines on the principles of good distribution practices for active substances for medicinal products for human use" (2015/C 95/01).

The guidelines set out the quality system elements for the procuring, importing, holding, supplying or exporting active substances. The scope of the guidelines exclude activities consisting of re-packaging, re-labelling or dividing up of active substances which are manufacturing activities and as such are subject to the guidelines on GMP of active substances. The guidelines, which can be found in Chapter 12 of this guide, cover:

- Quality System
- Personnel
- Documentation
- Orders
- Procedures
- Records
- Premises and Equipment
- Receipt
- Storage
- Deliveries to Customers
- Transfer of Information
- Returns
- Complaints and Recalls
- Self-inspections.

Written Confirmation

The Falsified Medicines Directive modifies EU Medicines Directive 2001/83/EC and from 2 July 2013 introduced new requirements for active substances imported into the EEA for use in the manufacture of authorised medicinal products.

The Falsified Medicines Directive requires importers of active substances to obtain written confirmations from competent authorities in non-EEA countries ("third countries") that the standards of manufacture of active substances at manufacturing sites on their territory are equivalent to EU good manufacturing practice (EU GMP). These confirmations are required before importation of active substances into the EU.

Each shipment of active substance received should be accompanied by a written confirmation from the competent authority of the exporting third country, stating that the active substance has been:

- manufactured to GMP standards at least equivalent to those laid down in the European Union;
- the third country manufacturing plant is subject to regular, strict and transparent inspections, and effective enforcement of GMP;
- in the event of non-conformance of the manufacturing site on inspection, such findings will be communicated to the European Union without delay.

The Commission's questions and answers document on this subject can be found on the Commission's website[2] and has been reproduced in Appendix 5 of this guide.

The template for the written confirmation has been published in Part III of EudraLex, Volume 4 and is reproduced here.

Template for the 'written confirmation' for active substances exported to the European Union for medicinal products for human use, in accordance with Article 46b(2)(b) of Directive 2001/83/EC

(1) Directive 2011/62/EU of the European Parliament and of the Council of 8 June 2011 amending Directive 2001/83/EC on the Community code relating to medicinal products for human use, as regards the prevention of the entry into the legal supply chain of falsified medicinal products (OJ L 174, 1.7.2011, p. 74) introduces EU-wide rules for the importation of active substances: According to Article 46b(2) of Directive 2001/83/EC, active substances shall only be imported if, *inter alia*, the active substances are accompanied by a written confirmation from the competent authority of the exporting third country which, as regards the plant manufacturing the exported active substance, confirms that the standards of good manufacturing practice and control of the plant are equivalent to those in the Union.

(2) The template for this written confirmation is set out in annex.

[2] http://ec.europa.eu/health/files/gmp/qa_importation.pdf

ANNEX

Letterhead of the issuing regulatory authority

Written confirmation for active substances exported to the European Union (EU) for medicinal products for human use, in accordance with Article 46b(2)(b) of Directive 2001/83/EC

Confirmation no. (given by the issuing regulatory authority):

....

1. Name and address of site (including building number, where applicable):

....

2. Manufacturer's licence number(s):[3]

....

Regarding the Manufacturing Plant under (1) of the Following Active Substance(s) Exported to the EU for Medicinal Products for Human Use

Active substance(s):[4]	Activity(ies):[5]

- **The Issuing Regulatory Authority Hereby Confirms That:**

The standards of good manufacturing practice (GMP) applicable to this manufacturing plant are at least equivalent to those laid down in the EU (= GMP of WHO/ICH Q7);

The manufacturing plant is subject to regular, strict and transparent controls and to the effective enforcement of good manufacturing practice, including repeated and unannounced inspections, so as to ensure a protection of public health at least equivalent to that in the EU; and

In the event of findings relating to non-compliance, information on such findings is supplied by the exporting third country without delay to the EU.[6]

[3] Where the regulatory authority issues a licence for the site. Record 'not applicable' in case where there is no legal framework for issuing of a licence.

[4] Identification of the specific active substances through an internationally-agreed terminology (preferably international nonproprietary name).

[5] For example, 'Chemical synthesis', 'Extraction from natural sources', 'Biological processes', 'Finishing steps'.

[6] qdefect@ema.europa.eu.

Date of inspection of the plant under (1). Name of inspecting authority if different from the issuing regulatory authority:

....

This written confirmation remains valid until

....

The authenticity of this written confirmation may be verified with the issuing regulatory authority.

This written confirmation is without prejudice to the responsibilities of the manufacturer to ensure the quality of the medicinal product in accordance with Directive 2001/83/EC.

Address of the issuing regulatory authority:

....

Name and function of responsible person:

....

E-mail, Telephone no., and Fax no.:

....

Signature Stamp of the authority and date

Waiver from Written Confirmation

The Falsified Medicines Directive provides two waivers if written confirmations are not provided. The first is where the regulatory framework applicable to active substances in those third countries has been assessed by the European Commission (EC) as providing an equivalent level of protection of public health over active substance manufacture and distribution to those applied in the EU. This assessment follows a request from the exporting third country's competent authority, and considers the regulatory framework for active substance manufacture and control and its equivalence to EU standards. Only if the exporting country is on the EC's ("white") list[7] is the requirement for a written confirmation from that country's competent authority removed.

The second waiver is where the third country active substance manufacturing site has been inspected by an EU Member State, has issued

[7] http://ec.europa.eu/health/human-use/quality/index_en.htm

a certificate of compliance with EU-GMP and that it remains within its period of validity. This is an exceptional waiver intended to apply where it is necessary to ensure the availability of medicinal products. Member States using this waiver should communicate this fact to the European Commission in accordance with the legislation (Article 46b(4)). MHRA has notified the European Commission of its intent to use this waiver should that be necessary.

Procedure for Active Substance Importation

A summary of the overall active substance import process has been developed to promote a common expectation and common approaches by Member States and is available on the Heads of Medicines Agencies, website: http://www.hma.eu/43.html. The flow chart is reproduced here:

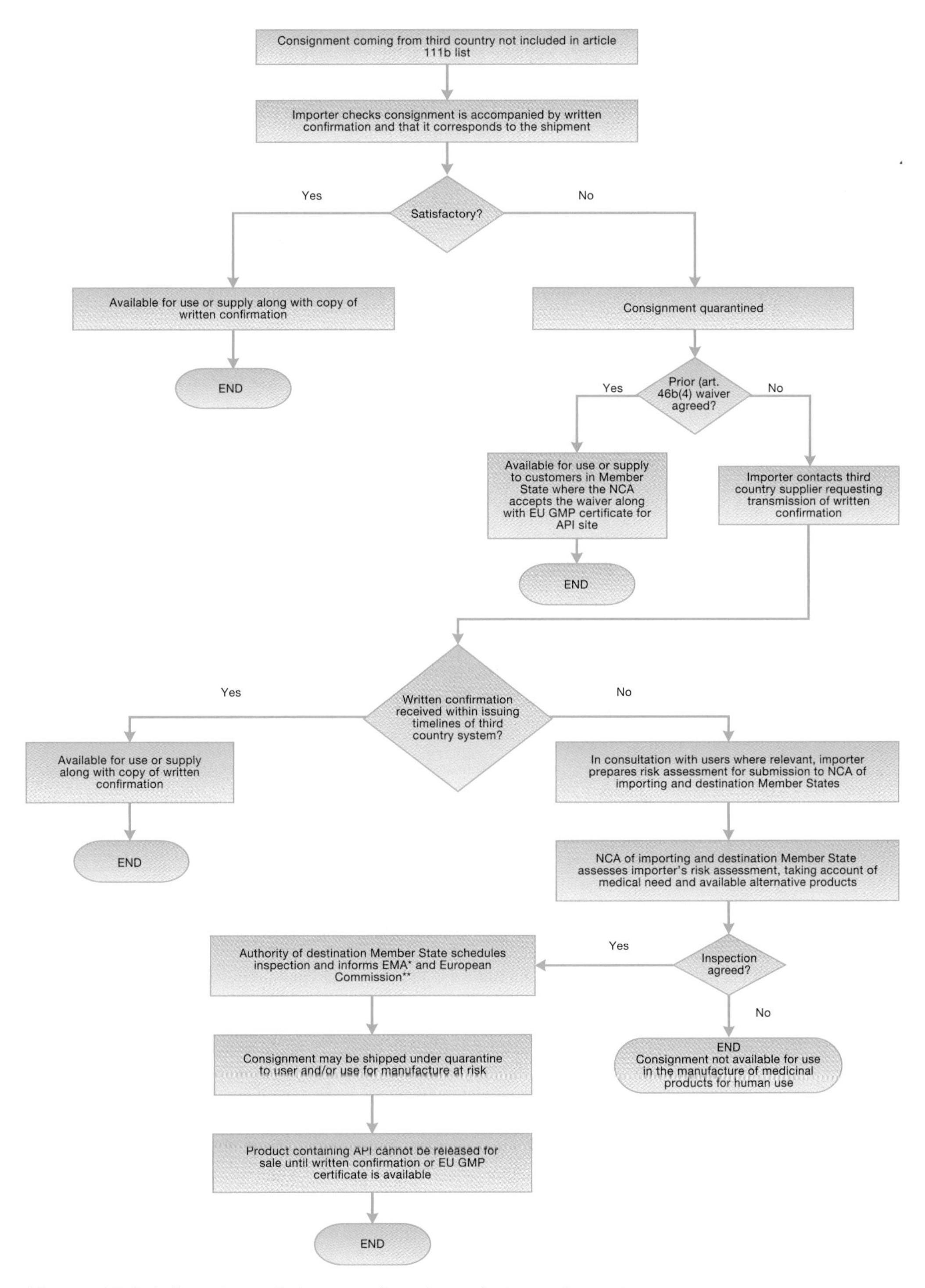

Figure 13.2 A flowchart of the overall active substance import process

*GMPINS@ema.europa.eu

**sanco-pharmaceuticals-d6@ec.europa.eu

Procedure for Waiver from Written Confirmation

UK based companies (registered importers and manufacturers who also import directly) who wish to import active substances under the second waiver should apply to gmpinspectorate@mhra.gsi.gov.uk using the form provided below which is available on the MHRA website.

Application Form for a waiver from the requirement to supply Written Confirmation with consignments of an imported Active Substance on the basis of a GMP certification of the Active Substance Manufacturer by an EEA Member State

A. Details of Third Country Manufacturing Site

Name of Active Substance :____________________

(Note: Only one active substance permitted per Waiver application. Use INN nomenclature)

Name of Active Substance Manufacturer :____________________

Address of Active Substance Manufacturer :____________________

____

____

Country :____________________

Third Country Competent Authority :____________________
site / facility reference number (if known)

B. Reason for Application for this Waiver

The manufacturer / importer should attach a document explaining the reason for requesting this waiver. It should be noted that if the active substance is being sourced from a third country where the authority there is known to issue Written Confirmations then under normal circumstances it would be expected that a Written Confirmation would be the basis for importation of active substances from that country.

C. Details of GMP Certification

Any differences between the name and address supplied above and those details supplied on the GMP certificate must be justified in order to ensure efficient processing of the application.

Name of authority which issued the GMP certificate :____________________

Inspection Date Referenced on GMP certificate :____________________

Period of validity of GMP certificate* (if stated) :____________________

(*this is 3 years unless there is a statement to the contrary)

Please attach a copy of the GMP certificate to this application form.

D. Details of Waiver Applicant

The Waiver Application may be submitted either by a site which has been registered with the MHRA for importation activities in the UK relating to active substances or by an authorised manufacturer / importer of human medicines in the UK which is using the imported active substance for manufacture of medicines for human use (excludes Investigational Medicinal Products) at its manufacturing address. If the importation activities (purchase of the active substance from a third country site or acting as the direct site of physical importation of the active substance) are being carried out directly by an authorised manufacturer of human medicines then both sections D1 and D2 below should be completed.

D1. Active Substance Importer

Active Substance Importers Registration No.:____________________

Importation activity carried out for this active substance
(tick all that apply) ☐ Procurement (Purchasing) ☐ Site of physical importation

Registered Name of Importer :____________________

Registered Address of Importer :____________________________________

Country UK

D2. Authorised Manufacturer of Human Medicines using Imported Active Substance
Manufacturer's / Importers Authorisation No. : ____________________
Name of Authorised Manufacturer : ____________________
Manufacturing Site Address : ____________________

Country UK

E. Signature of Waiver Applicant
Signature :____________________ Date:__________
Name (Print) :____________________
Position :____________________

F. Decision by MHRA on Waiver
Waiver Number:____________________
Waiver approved (Yes / No):____________________
Signature:____________________ Date:__________
Name (Print):____________________
Position:____________________

National Contingency Guidance

MHRA acknowledges that these requirements may in some instances make the sourcing of active substances from some third countries difficult. MHRA has therefore developed contingency measures that would allow the Agency, in cases where there is an overriding need to ensure continued supply of specific active substances, to provide an opinion on the importation of the active substance to permit manufacture, QP certification and supply of finished medicinal products. The aim of these contingency measures is to ensure, as per the aims of the Falsified Medicines Directive, the continued supply of active substances of appropriate quality and maintain the responsibility for the quality of the authorised medicinal products with the manufacturer, as is the case under the current legislation in force. Where these supply difficulties exist, please contact the GMP inspectorate at GMPinspectorate@mhra.gsi.gov.uk.

These measures may be used where:

- the third country active substance manufacturing site is not covered by a written confirmation; and
- the exporting country has not been assessed by the EC as having standards equivalent to EU GMP; and
- the third country active substance manufacturing site is not the subject, following inspection, of a current certificate of compliance with EU GMP,

Where UK Manufacturing Authorisation Holders ("MIA Holder") have determined that their third country active substance sources are at risk and have not been contacted by MHRA they should provide evidence in a submission / declaration template that:

- the active substance manufacturing site has been audited in the last three years either by himself or by a third party acting on his behalf and found to be operating in compliance with EU GMP for active substances; and
- the third country active substance manufacturing site is the subject, following inspection, of a current certificate of compliance with GMP issued by a recognised national authority or international organisation e.g. US-FDA, EU-MRA partners, EU-ACCA partners, PIC/S member states and the WHO.

UK based companies (registered importers and manufacturers who also import directly) who wish to import active substances under this guidance should apply to GMPinspectorate@mhra.gsi.gov.uk using the form provided at the end of this section which is available on the MHRA website.

Where the MIA Holder can make the first declaration but cannot declare that the third country active substance manufacturing site is the subject, following inspection, of a current certificate of compliance with GMP issued by a recognised national authority or international organisation then MHRA intends to enter details of the third country active substance manufacturing site onto a database of pending GMP inspection of a third country active substance manufacturing site.

MHRA will conduct further assessment of the data supplied in the submission(s) at the next routine reinspection of these finished product manufacturing sites.

The entry on the database will be removed if the active substance manufacturing site and the AS subsequently become the subject of a written confirmation, the third country has been assessed as having standards equivalent to GMP, or the AS manufacturing site, following inspection has been issued with a certificate of compliance with EU GMP. The pending-inspection database will only be available as long as this is required as a contingency measure.

Importers of active substances are also asked to note that where the sourcing of active substances from some third countries is difficult because it is not covered by a written confirmation, the exporting country has not been assessed as equivalent and the active substance manufacturing site is not the subject of a valid GMP certificate, these circumstances are subject to on-going review coordinated at an EU level. A key element of this review is the gathering of further data from EU-based finished product manufacturers for active substance import risk assessment and, where required, the EU level coordination of third country active substance manufacturer inspections.

National Contingency Guidance Submission Template

PART A: Finished Product Manufacturer to which this declaration applies

Name and address of MIA holder	Authorisation number

PART B: Concerned Third Country Active Substance Manufacturing Site

Name and Address of Active Substance Manufacturing Site
Name of Active Substances manufactured at this site

PART C: Basis of declaration in lieu of full compliance with Article 46b(2)

Please tick to confirm that an on-site audit of the active substance manufacturer has been conducted by the MIA holder or by a third party on their behalf

(i) ☐ **On-site audit of the active substance manufacturer(s) conducted by MIA holder or by a third party on their behalf:**

An on site audit of the active substance(s) manufactured at the site listed in PART B has been completed either by the MIAH(s) listed below or by a third party auditing body i.e. contract acceptor(s) on behalf of the MIAH i.e. contract giver(s) as listed:

MIAH (or contract giver)	Auditing body (contract acceptor)	Site audited	Date of audit[1]

[1]The date of the last audit should not exceed 3 years.

(ii) **Availability of a current[2] certificate of compliance with GMP issued by a recognised national authority or international organisation e.g. EDQM-CEP, US-FDA, EU-MRA partners, EU-ACCA partners, PIC/S member states and the WHO.**

☐ A current certificate for the site named in Section B of compliance with GMP issued by a recognised national authority or international organisation **is** available.

☐ A current certificate for the site named in Section B of compliance with GMP issued by a recognised national authority or international organisation **is not** available.

Inspection Authority	Date of inspection

[2]The date of the last inspection should not exceed 3 years.

(iii) **Supplementary supportive information (optional):**

For the active substance manufacturing site listed, results of inspection report(s) or GMP certificate(s) issued by recognised national authorities or international organisations together with other supporting information are attached.

Summary of supporting information provided

PART D: Declaration

I declare that:
QP Responsibility

- I am a QP with specific responsibility for GMP compliance of the active substance manufactured at the sites listed in Part B and am authorised to make this declaration.
- That the audit report(s) and all the other documentation relating to this declaration of GMP compliance of the active substance manufacturer(s) will be made available for inspection by the competent authorities, if requested.

GMP Compliance

- The manufacture of the named active substance(s) at the site given in Part B is in accordance with the detailed guideline on good manufacturing practice for active substances used as starting materials as required by Article 46(f) of Directive 2001/83/EC as amended.
- This is based upon an on-site audit of the active substance manufacturer.
- That the outcome of the audit confirms that the active substance manufacturer complies with the principles and guidelines of good manufacturing practice.

Audit

- In the case of third party audit(s), I have evaluated each of the named contract acceptor(s) give in Part C and that technical contractual arrangements are in place and that any measures taken by the contract giver(s) are documented e.g. signed undertakings by the auditor(s).
- In all cases, the audit(s) was/were conducted by properly qualified and trained staff, in accordance with approved procedures.

Inspection, of the third country AS manufacturing site by a recognised national authority or international organisation

- Where available results of inspection report(s) or GMP certificate(s) issued by recognised national authorities or international organisations are within their period of validity and are attached together with other supporting information.

PART E: Name and Signature of QP responsible for this Declaration

This declaration is submitted by:

Signatory ____________	Status (job title) ____________
Print name ____________	MIAH name: ____________
Date ____________	MIAH number: ____________

Appendices

Appendix 1

HUMAN AND VETERINARY MEDICINES AUTHORITIES IN EUROPE

Austria

Austrian Medicines and Medical Devices Agency – Austrian Federal Office for Safety in Health Care
Traisengasse 5
A-1200 Vienna
Austria
Phone +43 50 555 36111
Email BASG-AGESMedizinmarktaufsicht@ages.at
Website(s) http://www.basg.gv.at

Belgium

DG Enterprise and Industry – F/2 BREY 10/073
Avenue d'Auderghem 45
B-1049 Brussels
Belgium

Federal Agency for Medicines and Health Products
EUROSTATION
Victor Horta
40 40
1060 Brussels
Belgium
Email welcome@fagg-afmps.be
Website(s) http://www.fagg-afmps.be/en/

Bulgaria

Bulgarian Drug Agency
8, Damyan Gruev Str
1303 Sofia
Bulgaria
Phone +359 2 8903 555
Fax +359 2 8903 434
Email bda@bda.bg
Website(s) http://www.bda.bg

National Veterinary Service
15A, Pencho Slaveiko Blvd
1606 Sofia
Bulgaria
Phone +359 2 915 98 20
Fax +359 2 954 95 93
Email cvo@nvms.government.bg
Website(s) http://www.nvms.government.bg

Croatia

Croatian Agency for Medicinal Products and Medical Devices
Ksaverska c.4
10000 Zagreb
Croatia
Phone +385 1 4884 100
Fax +385 1 4884 110
Email halmed@halmed.hr
Website(s) http://www.halmed.hr

Ministry of Agriculture, Veterinary and Food Safety Directorate
Planinska Street 2a
10000 Zagreb
Croatia
Phone +385 6443 540
Fax +385 6443 899
Email veterinarstvo@mps.hr
Website(s) http://www.veterinarstvo.hr

Cyprus

Ministry of Health Pharmaceutical Services
7 Larnacos Ave
CY-1475 Nicosia
Cyprus
Website(s) http://www.moh.gov.cy

Czech Republic

State Institute for Drug Control
Srobárova 48
CZ-100 41 Praha 10
Czech Republic
Phone +420 272 185 111
Fax +420 271 732 377
Email posta@sukl.cz
Website(s) http://www.sukl.eu

Institute for State Control of Veterinary Biologicals and Medicaments
Hudcova Str. 56A
CZ-621 00 Brno-Medlánky
Czech Republic
Phone +420 541 518 211
Fax +420 541 210 026
Email uskvbl@uskvbl.cz
Website(s) http://www.uskvbl.cz/

Denmark

Danish Health and Medicines Agency
Axel Heides Gade 1
DK–2300 København S
Denmark
Phone +45 72 22 74 00
Fax +45 44 88 95 99
Email sst@sst.dk
Website(s) http://www.dkma.dk

European Commission

DG Enterprise F2 Pharmaceuticals
Rue de la Loi 200
B-1049 Brussels
Belgium

DG Health and Consumers: Unit D6: Medicinal Products – Quality, Safety and Efficacy; Unit D5: Medicinal Products – Authorisations, European Medicines Agency
Rue de la Loi 200
B-1049 Brussels
Belgium
Phone +32 2 299 11 11
Email: sanco-pharmaceuticals-d5@ec.europa.eu; sanco-pharmaceuticals-d6@ec.europa.eu
Website(s) http://ec.europa.eu/health/human-use/index_en.htm

European Medicines Agency (EMA)
7 Westferry Circus
London E14 4HB
United Kingdom
Phone +44 20 74 18 84 00
Fax +44 20 74 18 84 16
Email info@ema.europa.eu
Website(s) http://www.ema.europa.eu/

Estonia

State Agency of Medicines
1 Nooruse St
EE-50411 Tartu
Estonia
Phone +372 737 41 40
Fax +372 737 41 42
Email sam@sam.ee
Website(s) http://www.sam.ee/

Finland

Finnish Medicines Agency
PO Box 55
FI-00034 Fimea
Finland
Phone +358 29 522 3341
Fax +358 29 522 3001
Website(s) http://www.fimea.fi/

France

French National Agency for Medicines and Health Products Safety (ANSM)
143-147 bd
Anatole France
FR-93285 Saint Denis Cedex
France
Email ANSM@ansm.sante.fr
Website(s) http://www.ansm.sante.fr

Agence Nationale du Médicament Vétérinaire, Agence nationale de sécurité sanitaire de l'alimentation, de l'environnement et du travail (ANMV)
8 rue Claude Bourgelat, Parc d'Activités de la Grande Marche, Javené BP
CS 70611-35306 Fougères
France
Phone +33 (0) 2 99 94 78 71

Email sylvie.goby@anses.fr
Website(s) http://www.anses.fr

Germany

Federal Institute for Drugs and Medical Devices (BfArM)
Kurt-Georg-Kiesinger-Allee 3
53175 Bonn
Germany
Phone +49 (0)228-207-30
Fax +49 (0)228-207-5207
Email poststelle@bfarm.de
Website(s) http://www.bfarm.de

Federal Office of Consumer Protection and Food Safety
Mauerstraße 39–42
D-10117 Berlin
Germany
Phone +49 30 1 84 44-000
Fax +49 30 1 84 44-89 999
Email poststelle@bvl.bund.de
Website(s) http://www.bvl.bund.de (pharmaceuticals)

Paul-Ehrlich Institut (Federal Institute for Vaccines and Biomedicines)
Paul-Ehrlich-Straße 51–59
63225 Langen
Germany
Phone +49 6103 77 0
Fax +49 6103 77 1234
Email pei@pei.de
Website(s) http://www.pei.de
(vaccines, blood products, sera)

Greece

National Organization for Medicines
Messogion Avenue 284, cholargos
GR-15562 Athens
Greece
Phone +30 213 2040 3846507200
Fax +30 213 2040 569 6545535
Email inter.rel@eof.gr
Website(s) http://www.eof.gr/
(pharmaceuticals and immunologicals)

Hungary

National Institute of Pharmacy and Nutrition
Zrínyi U. 3
H-1051 Budapest
Hungary
Website(s) http://www.ogyei.hu/

National Food Chain Safety Office, Directorate of Veterinary Medicinal Products
Szállás utca 8
H-1107 Budapest 10.Pf. 318
Hungary
Phone +36 1 433 03 30
Fax +36 1 262 28 39
Email info.aogyti@oai.hu
Website(s) http://www.nebih.gov.hu/en/specialities/veterinary

Iceland

Icelandic Medicines Agency
Vinlandsleid 14
IS-113 Reykjavik
Iceland
Phone +354 520 2100
Fax +354 561 2170
Email ima@ima.is
Website(s) http://www.ima.is

Ireland

Health Products Regulatory Agency (HPRA)
Kevin O'Malley House
Earlsfort Centre
Earlsfort Terrace
IRL-Dublin 2
Ireland
Phone +353 1 676 4971
Fax +353 1 676 7836
Email info@hpra.ie
Website(s) http://www.hpra.ie

Department of Agriculture and Food
Kildare St
Dublin
Ireland
Phone +353 1 607 20 00
Fax +353 1 661 62 63
Email info@agriculture.gov.ie
Website(s) http://www.agriculture.gov.ie

Italy

Italian Medicines Agency
Via del Tritone, 181
I-00187 Rome
Italy
Phone +39 6 5978401
Fax +39 6 59944142
Email forenamefirstletter.surename@aifa.gov.it
Website(s) http://www.agenziafarmaco.it/

Laboratorio di Medicina Veterinaria, Istituto Superiore di Sanità
Viale Regina Elena 299
I-00161 Rome
Italy
Phone +39 6 49 38 70 76
Fax +39 6 49 38 70 77

Ministero della Salute, Direzione Generale della Sanità Pubblica Veterinaria, degli Allimenti e della Nutrizione, Uff. XI
Piazzale G. Marconi 25
I-00144 Rome
Italy
Phone +39 06 59 94 65 84
Fax +39 06 59 94 69 49
Website(s) http://www.ministerosalute.it/

Latvia

State Agency of Medicines of Latvia
15 Jersikas Street
LV-1003 Riga
Latvia
Phone +371-7078424
Fax +371-7078428
Email info@zva.gov.lv
Website(s) http://www.zva.gov.lv/

Food and Veterinary Service
Peldu St 30
LV-1050 Riga
Latvia
Phone +371 67095271
Fax +371 67095270
Email nrd@pvd.gov.lv
Website(s) http://www.pvd.gov.lv

Liechtenstein

Office of Health/Medicinal Products Control Agency
Äulestr 512
FL-9490 Vaduz
Liechtenstein
Fax +423 236 73 50
Email pharminfo@llv.li
Website(s) www.llv.li/

Lithuania

State Medicines Control Agency
Žirmūnų str. 139A
LT-09120 Vilnius
Lithuania
Phone +370 5 263 92 64
Fax +370 5 263 92 65
Email vvkt@vvkt.lt
Website(s) http://www.vvkt.lt

National Food and Veterinary Risk Assessment Institute
J. Kairiukscio str. 10
LT-08409 Vilnius
Lithuania
Phone +370 5 2780470
Fax +370 5 2780471
Email nmvrvi@vet.lt
Website(s) http://www.nmvrvi.lt

State Food and Veterinary Service
Siesiku str. 19
LT-07170 Vilnius
Lithuania
Email vvt@vet.lt
Website(s) http://www.vet.lt

Luxembourg

Ministry of Health
Allée Marconi
L-2120 Luxembourg
Luxembourg
Phone +352 24785593
Fax +352 24795615
Email jacqueline.genoux-hames@ms.etat.lu
Email vet marc.schmit@ms.etat.lu
Website(s) http://www.ms.etat.lu

Malta

Medicines Authority
Sir Temi Zammit Buildings,

Malta Life Sciences Park
SGN 3000 San Gwann
Malta
Phone +356 23439000
Fax +356 23439161
Email info.medicinesauthority@gov.mt
Website(s) www.medicinesauthority.gov.mt

Veterinary Regulation Directorate
Abattoir Square
Albert Town
MRS 1123 Marsa
Malta
Phone +356 22925375
Email hmav.malta@gov.mt
Website(s) http://vafd.gov.mt/approved-vet-pharmaceaticals-in-malta

Netherlands

Medicines Evaluation Board
Graadt van Roggenweg 500
NL-3531 AH Utrecht
The Netherlands
Phone +31 (0) 88 - 224 80 00
Fax +31 (0) 88 - 224 80 01
Website(s) http://www.cbg-meb.nl/

Norway

Norwegian Medicines Agency (NOMA)
Sven Oftedalsvei 8
N-0950 Oslo
Norway
Phone +47 22 89 77 00
Fax +47 22 89 77 99
Email post@legemiddelverket.no
Website(s) http://www.legemiddelverket.no

Poland

Office for Registration of Medicinal Products, Medical Devices and Biocidal Products
Al.Jerozolimskie 181C
02-222 Warsaw
Poland
Phone +48 (22) 492 11 00
Fax +48 (22) 492 11 09
Website(s) http://bip.urpl.gov.pl

Portugal

INFARMED - National Authority of Medicines and Health Products, IP
Av. do Brasil 53
P-1749-004 Lisbon
Portugal
Phone +351 217987100
Fax +351 217987316
Email infarmed@infarmed.pt
Website(s) www.anm.ro/anmdm/en/index.html

Food and Veterinary Directorate General
Largo da Academia Nacional de Belas Artes, n. 2
1294-105 Lisbon
Portugal
Phone +351 21 323 95 00
Fax +351 21 346 35 18
Email dirgeral@dgav.pt
Website(s) http://www.anm.ro/anmdm/en/index.html

Romania

National Agency for Medicines and Medical Devices
48, Av. Sanatescu
011478 Bucharest
Romania
Phone +4021 3171100
Fax +4021 3163497
Website(s) http://www.anm.ro/anmdm/en/index.html

Institute for Control of Biological Products and Veterinary Medicines
Str. Dudului 37, sector 6
060603 Bucharest
Romania
Phone +40 21 220 21 12
Fax +40 21 221 31 71
Email icbmv@icbmv.ro
Website(s) http://www.icbmv.ro/

Slovakia

State Institute for Drug Control
Kvetná 11
SK-825 08 Bratislava 26
Slovakia
Phone +421 2 5070 1111

Fax +421 2 5556 4127
Email sukl@sukl.sk
Website(s) http://www.sukl.sk/

Institute for State Control of Veterinary Biologicals and Medicaments
Biovetská 34
PO Box 52c
SK-949 01 Nitra
Slovakia
Phone +421 37 6515506
Fax +421 37 6517915
Email uskvbl@uskvbl.sk
Website(s) http://www.uskvbl.sk/

Slovenia

Javna agencija Republike Slovenije za zdravila in medicinske pripomočke
Ptujska ulica 21
Sl-1000 Ljubljana
Slovenia
Phone +38 6 8 2000 500
Fax +38 6 8 2000 510
Email info@jazmp.si
Website(s) http://www.jazmp.si

Spain

Spanish Agency of Medicines and Medical Devices
Parque Empresarial Las
Mercedes Edificio 8C
Campezo, 1
E-28022 Madrid
Spain
Phone +34 91 8225997
Fax +34 91 8225128
Email internacional@aemps.es
Website(s) www.agemed.es

Sweden

Medical Products Agency
Dag Hammarskjölds väg 42 / Box 26
SE-751 03 Uppsala
Sweden
Phone +46 (0) 18 17 46 00
Fax +46 (0) 18 54 85 66
Email registrator@mpa.se
Website(s) www.lakemedelsverket.se

United Kingdom

Medicines and Healthcare products Regulatory Agency
151 Buckingham Palace Road
Victoria
London
SW1W 9SZ
United Kingdom
Phone +44 (0)20 3080 6000
Fax +44 (0)203 118 9803
Website(s) www.mhra.gov.uk

Veterinary Medicines Directorate (VMD)
Woodham Lane
New Haw
Addlestone
KT15 3LS
United Kingdom
Phone +44 1932 33 69 11
Fax +44 1932 33 66 18
Email postmaster@vmd.defra.gsi.gov.uk
Website(s) www.vmd.defra.gov.uk

Appendix 2

LIST OF PERSONS WHO CAN BE SUPPLIED WITH MEDICINES BY WAY OF WHOLESALE DEALING (HUMAN MEDICINES REGULATIONS 2012)

Contents

All Medicines

- Doctors and dentists[1]
- Persons lawfully conducting a retail pharmacy business within the meaning of section 69 of the Medicines Act 1968
- Holders of wholesale dealer's licences or persons to whom the restrictions imposed by regulation 18(1) of the Human Medicines Regulations do not apply because of an exemption in those Regulations
- Authorities or persons carrying on the business of -
  - (a) an independent hospital, independent clinic or independent medical agency, or
  - (b) a hospital or health centre which is not an independent hospital or independent clinic
  - (c) a nursing home in Northern Ireland
- Ministers of the Crown and Government Departments
- Scottish Ministers
- Welsh Ministers

[1] Health professionals may order medicines using a doctor or dentist's letter of authority. The letter should be clear that the health professional is authorised to order medicines on the doctor or dentist's behalf for delivery to that doctor or dentist.

- Northern Ireland Ministers
- A person other than an excepted person (in this context a pharmacist) who carries on a business consisting (wholly or partly) of the supply or administration of medicinal products for the purpose of assisting the provision of health care by or on behalf of, or under arrangements made by –
  (a) a police force in England, Wales or Scotland,
  (b) the Police Service of Northern Ireland,
  (c) a prison service, or
  (d) Her Majesty's Forces.
- A person other than an excepted person (in this context a pharmacist) who carries on a business consisting (wholly or partly) of the supply or administration of medicinal products for the purpose of assisting the provision of health care by or on behalf of, or under arrangements made by –
  (a) NHS trust or NHS foundation trust
  (b) The Common Services Agency
  (c) A local authority in the exercise of public health functions (within the meaning of the NHS Act 2006)
  (d) A health authority or special health authority
  (e) A clinical commissioning group
  (f) NHS England
  (g) A local authority

Pharmacy Medicines

- Pharmacy medicines which are for the purpose of administration in the course of a business can be supplied to a person carrying on that business.

General Sale List Medicines

- Any person who requires GSL medicines for the purpose of selling, supplying or administering them to human beings in the course of a business carried on by that person.

Other Relevant Provisions

- Under medicines legislation, the general rule is that prescription only medicines may only be sold or supplied on a retail basis against a prescription on registered pharmacy premises by or under the supervision of a pharmacist. There are exemptions from these restrictions for certain persons in respect of specific medicines which are contained in

Schedule 17 to the Human Medicines Regulations 2012. The Regulations allow people who have exemptions to be sold the prescription only medicines relevant to their particular exemption on a wholesale basis. **See Schedule 17.**

- Similarly, the general rule for pharmacy medicines is that they may only be sold or supplied by or under the supervision of a pharmacist on registered pharmacy premises. Again, there are exemptions from these restrictions for certain persons in respect of specific medicines contained in Schedule 17 of the Human Medicines Regulations 2012 and the Regulations allow for the sale of the P medicines on the same basis as for POMs (above). **See Schedule 17.**

Additional Provisions for Optometrists and Dispensing Opticians not included in Schedule 17

- The Human Medicines Regulations also contain other provisions which allow wholesale sales in certain circumstances:
  (**a**) Registered optometrists can obtain a range of medicines on a wholesale basis for sale or supply in the course of their professional practice. In addition to these arrangements, they may also obtain prescription only medicines containing the following substances for administration:
  Amethocaine hydrochloride
  Lidocaine hydrochloride
  Oxybuprocaine hydrochloride
  Proxymetacaine hydrochloride
  (**b**) Additional Supply Optometrists can be sold a product that is only classed as a POM because it contains Thymoxamine Hydrochloride
  (**c**) Dispensing opticians can obtain the following which are required for use by a registered optometrist or doctor attending their practice:
  Amethocaine Hydrochloride
  Chloramphenicol
  Cyclopentolate Hydrochloride
  Fusidic Acid
  Lidocaine Hydrochloride
  Oxybuprocaine Hydrochloride
  Proxymetacaine Hydrochloride
  Tropicamide
  (**d**) Dispensing opticians can also obtain medicines that are required in the course of their professional practice as a contact lens specialist which contain one or more of the following substances:
  Lidocaine Hydrochloride
  Oxybuprocaine Hydrochloride
  Proxymetacaine Hydrochloride

SCHEDULE 17 PROVISIONS

PART 1 Exemption from restrictions on Sale or Supply of Prescription Only Medicines

Persons exempted	Prescription only medicines to which the exemption applies	Conditions
1. Persons selling or supplying prescription only medicines to universities, other institutions concerned with higher education or institutions concerned with research.	1. All prescription only medicines.	1. The sale or supply shall be— (a) subject to the presentation of an order signed by the principal of an institution concerned with educational research or the appropriate head of department in charge of a specified course of research stating— (i) the name of the institution for which the prescription only medicine is required, and (ii) the purpose for which the prescription only medicine is required, and (iii) the total quantity required; and (b) for the purpose of the education or research with which the institution is concerned.
2. Persons selling or supplying prescription only medicines to any of the following— (a) a public analyst appointed under section 27 of the Food Safety Act 1990(1) or article 27 of the Food Safety (Northern Ireland) Order 1991(2); (b) an authorised officer within the meaning of section 5(6) of the Food Safety Act 1990(3); (c) a sampling officer within the meaning of article 38(1) of the Food (Northern Ireland) Order 1989(4); (d) an inspector acting under regulations 325 to 328; (e) a sampling officer within the meaning of Schedule 31.	2. All prescription only medicines.	2. The sale or supply shall be subject to the presentation of an order signed by or on behalf of any person listed in column 1 stating the status of the person signing it and the amount of prescription only medicine required, and shall be only in connection with the e xercise by those persons of their statutory functions.
3. Persons selling or supplying prescription only medicines to any person employed or engaged in connection with a scheme for testing the quality and checking the amount of the drugs and appliances supplied under the National Health Service Act 2006(5), the National Health Service (Scotland) Act	3. All prescription only medicines.	3. The sale or supply shall be— (a) subject to the presentation of an order signed by or on behalf of the person so employed or engaged stating the status of the person signing it and the amount of the prescription only medicine required; and

Persons exempted	Prescription only medicines to which the exemption applies	Conditions
1978(6), the National Health Service (Wales) Act 2006(7) and the Health and Personal Social Services (Northern Ireland) Order 1972(8), or under any subordinate legislation made under those Acts or that Order.		(b) for the purposes of a scheme referred to in column 1 in this paragraph.
4. Registered midwives.	4. Prescription only medicines containing any of the following substances— (a) Diclofenac; (b) Hydrocortisone Acetate; (c) Miconazole; (d) Nystatin; (e) Phytomenadione;	4. The sale or supply shall be only in the course of their professional practice.
5. Persons lawfully conducting a retail pharmacy business within the meaning of section 69 of the Medicines Act 1968.	5. Water for injection.	5. The sale or supply is to a person— (a) for a purpose other than parenteral administration; or (b) who has been prescribed dry powder for parenteral administration but has not been prescribed the water for injection that is needed as a diluent.
6. Persons lawfully conducting a retail pharmacy business within the meaning of section 69 of the Medicines Act 1968.	6. Items which are— (a) prescription only medicines which are not for parenteral administration and which— (i) are eye drops and are prescription only medicines by reason only that they contain not more than 0.5 per cent of Chloramphenicol, or (ii) are eye ointments and are prescription only medicines by reason only that they contain not more than 1.0 per cent Chloramphenicol, or (iii) are prescription only medicines by reason only that they contain any of the following substances— (aa) Cyclopentolate hydrochloride, (bb) Fusidic Acid, (cc) Tropicamide; (b) the following prescription only medicines—	6. The sale or supply shall be subject to the presentation of an order signed by— (a) a registered optometrist for a medicine listed under item (a) in column 2; (b) a registered chiropodist or podiatrist for a medicine listed under item (b) in column 2.

Persons exempted	Prescription only medicines to which the exemption applies	Conditions
	(i) Amorolfine hydrochloride cream where the maximum strength of the Amorolfine in the cream does not exceed 0.25 per cent by weight in weight, (ii) Amorolfine hydrochloride lacquer where the maximum strength of Amorolfine in lacquer does not exceed 5 per cent by weight in volume, (iii) Amoxicillin, (iv) Co-Codamol, (v) Co-dydramol 10/500 tablets, (vi) Codeine Phosphate, (vii) Erythromycin, (viii) Flucloxacillin, (ix) Silver Sulfadiazine, (x) Tioconazole 28%, (xi) Topical hydrocortisone where the maximum strength of hydrocortisone in the medicinal product does not exceed 1 per cent by weight in weight.	
7. Registered optometrists.	7. Prescription only medicines listed in item (a) of paragraph 6 column 2.	7. The sale or supply shall be only— (a) in the course of their professional practice, and (b) in an emergency.
8. Persons lawfully conducting a retail pharmacy business within the meaning of section 69 of the Medicines Act 1968.	8. Medicinal products not for parenteral administration which are prescription only medicines by reason only that they contain any of the following substances— (a) Acetylcysteine, (b) Atropine sulphate, (c) Azelastine hydrochloride, (d) Diclofenac sodium, (e) Emedastine, (f) Homotropinehydrobromide, (g) Ketotifen, (h) Levocabastine, (i) Lodoxamide, (j) Nedocromil sodium,	8. The sale or supply shall be subject to the presentation of an order signed by an additional supply optometrist.

Persons exempted	Prescription only medicines to which the exemption applies	Conditions
	(k) Olopatadine, (l) Pilocarpine hydrochloride, (m) Pilocarpine nitrate, (n) Polymyxin B/bacitracin, (o) Polymyxin B/trimethoprim, (p) Sodium cromoglycate.	
9. Additional supply optometrists.	9. Prescription only medicines specified in paragraph 8 column 2.	9. The sale or supply shall be only— (a) in the course of their professional practice, and (b) in an emergency.
10. Holders of marketing authorisations, product licences or manufacturer's licences.	10. Prescription only medicines referred to in those authorisations or licences.	10. The sale or supply shall be only— (a) to a pharmacist, (b) so as to enable that pharmacist to prepare an entry relating to the prescription only medicine in question in a tablet or capsule identification guide or similar publication, and (c) of no greater quantity than is reasonably necessary for that purpose.
11. Registered chiropodists or podiatrists against whose names are recorded in the relevant register annotations signifying that they are qualified to use the medicine specified in column 2.	11. The following prescription only medicines— (a) Amorolfine hydrochloride cream where the maximum strength of the Amorolfine in the cream does not exceed 0.25 per cent by weight in weight, (b) Amorolfine hydrochloride lacquer where the maximum strength of Amorolfine in lacquer does not exceed 5 per cent by weight in volume, (c) Amoxicillin, (d) Co-Codamol, (e) Co-dydramol 10/500 tablets, (f) Codeine Phosphate, (g) Erythromycin, (h) Flucloxacillin, (i) Silver Sulfadiazine, (j) Tioconazole 28%, (k) Topical hydrocortisone where the maximum strength of hydrocortisone in the medicinal product does not exceed 1 per cent by weight in weight.	11. The sale or supply shall be only in the course of their professional practice.

Persons exempted	Prescription only medicines to which the exemption applies	Conditions
12. Persons selling or supplying prescription only medicines to a school.	12. A prescription only medicinal product comprising an inhaler containi ng salbutamol.	12. The sale or supply shall be— (a) subject to the presentation of an order signed by the principal or head teacher at the school concerned stating— (i) the name of the school for which the medicinal product is required, (ii) the purpose for which that product is required, and (iii) the total quantity required, and (b) for the purpose of supplying the medicinal product to pupils at the school in an emergency.
13. Registered orthoptists.	13. The following prescription only medicines— (a) Atropine, (b) Cyclopentolate, (c) Tropicamide, (d) Lidocaine with fluorescein, (e) Oxybuprocaine, (f) Proxymetacaine, (g) Tetracaine, (h) Chloramphenicol, (i) Fusidic acid.	13. The sale or supply shall be only in the course of their professional practice.

PART 2 Exemption from the restriction on supply of prescription only medicines

Persons exempted	Prescription only medicines to which the exemption applies	Conditions
1. Royal National Lifeboat Institution and certified first aiders of the Institution.	1. All prescription only medicines	1. The supply shall be only so far as is necessary for the treatment of sick or injured persons in the exercise of the functions of the Institution.
2. The owner or master of a ship which does not carry a doctor on board as part of the ship's complement.	2. All prescription only medicines.	2. The supply shall be only so far as is necessary for the treatment of persons on the ship.
3. Persons authorised by licences granted under regulation 5 of the Misuse of Drugs Regulations 2001(**9**) or regulation 5 of the Misuse of Drugs Regulations (Northern Ireland) 2002(**10**) to supply a controlled drug.	3. Such prescription only medicines, being controlled drugs, as are specified in the licence.	3. The supply shall be subject to such conditions and in such circumstances and to such an extent as may be specified in the licence.

Persons exempted	Prescription only medicines to which the exemption applies	Conditions
4. Persons employed or engaged in the provision of lawful drug treatment services.	4. Ampoules of sterile water for injection that contain no more than 2 ml of water each.	4. The supply shall be only in the course of provisions of lawful drug treatment services.
5. Persons requiring prescription only medicines for the purpose of enabling them, in the course of any business carried on by them, to comply with any requirements made by or in pursuance of any enactment with respect to the medical treatment of their employees.	5. Such prescription only medicines as may be specified in the relevant enactment.	5. The supply shall be— (a) for the purpose of enabling them to comply with any requirements made by or in pursuance of any such enactment, and (b) subject to such conditions and such circumstances as may be specified in the relevant enactment.
6. Persons operating an occupational health scheme.	6. Prescription only medicines sold or supplied to a person operating an occupational health scheme in response to an order in writing signed by a doctor or a registered nurse.	6. The supply of the prescription only medicine shall be— (a) in the course of operating an occupational health scheme, and (b) made by— (i) a doctor, or (ii) a registered nurse acting in accordance with the written directions of a doctor as to the circumstance in which such medicines are to be used in the course of an occupational health scheme.
7. The operator or commander of an aircraft.	7. Prescription only medicines which are not for parenteral administration and which have been sold or supplied to an operator or commander of an aircraft in response to an order in writing signed by a doctor.	7. The supply shall be only so far as is necessary for the immediate treatment of sick or injured persons on the aircraft and shall be in accordance with the written instructions of a doctor as to the circumstances in which prescription only medicines of the description in question are to be used on the aircraft.
8. Persons employed as qualified first-aid personnel on off-shore installations.	8. All prescription only medicines.	8. The supply shall be only so far as is necessary for the treatment of persons on the installation.
9. Persons who hold a certificate in first aid from the Mountain Rescue Council of England and Wales, or from the Northern Ireland Mountain Rescue Co-ordinating Committee.	9. Prescription only medicines supplied to a person specified in column 1 in response to an order in writing signed by a doctor.	9. The supply shall be only so far as is necessary for the treatment of sick or injured persons in the course of providing mountain rescue services.
10. Persons ("P") who are members of Her Majesty's armed forces.	10. All prescription only medicines.	10. The supply shall be— (a) in the course of P undertaking any function as a member of Her Majesty's armed forces; and (b) where P is satisfied that it is not practicable for another person who is legally entitled to supply a prescription only medicine to do so; and

Persons exempted	Prescription only medicines to which the exemption applies	Conditions
		(c) only in so far as is necessary— (i) for the treatment of a sick or injured person in a medical emergency, or (ii) to prevent ill-health where there is a risk that a person would suffer ill-health if the prescription only medicine is not supplied.
11. A person ("P") carrying on the business of a school who is trained to administer the relevant medicine.	11. A prescription only medicinal product comprising an inhaler containing salbutamol.	11. The supply shall be— (a) in the course of P carrying on the business of a school; (b) where supply is to a pupil at that school who is known to suffer from asthma; and (c) where the pupil requires the medicinal product in an emergency.
12. Registered midwives.	12. Prescription only medicines for parenteral administration that contain– (a) Diamorphine, (b) Morphine, (c) Pethidine hydrochloride.	12. The supply shall be only in the course of their professional practice.

PART 3 Exemptions from the restriction on administration of prescription only medicines

Persons exempted	Prescription only medicines to which the exemption applies	Conditions
1. Registered chiropodists or podiatrists against whose names are recorded in the relevant register annotations signifying that they are qualified to use the medicines specified in column 2.	1. Prescription only medicines for parenteral administration that contain— (a) Adrenaline, (b) Bupivacaine hydrochloride, (c) Bupivacaine hydrochloride with adrenaline where the maximum strength of adrenaline does not exceed 1 mg in 200 ml of bupivacaine hydrochloride, (d) Levobupivacaine hydrochloride, (e) Lidocaine hydrochloride, (f) Lidocaine hydrochloride with adrenaline where the maximum strength of adrenaline does not exceed 1 mg in 200 ml of lignocaine hydrochloride, (g) Mepivacaine hydrochloride, (h) Methylprednisolone, (i) Prilocaine hydrochloride, (j) Ropivacaine hydrochloride.	1. The administration shall only be in the course of their professional practice and where the medicine includes a combination of substances in column 2, those substances shall not have been combined by the chiropodist or podiatrist.

Persons exempted	Prescription only medicines to which the exemption applies	Conditions
2. Registered midwives and student midwives.	2. Prescription only medicines for parenteral administration containing any of the following substances but no other substance that is classified as a product available on prescription only— (a) Adrenaline, (b) Anti-D immunoglobulin, (c) Carboprost, (d) Cyclizine lactate, (e) Diamorphine, (f) Ergometrine maleate, (g) Gelofusine, (h) Hartmann's solution, (i) Hepatitis B vaccine, (j) Hepatitis immunoglobulin, (k) Lidocaine hydrochloride, (l) Morphine, (m) Naloxone hydrochloride, (n) Oxytocins, natural and synthetic, (o) Pethidine hydrochloride, (p) Phytomenadione, (q) Prochloperazine, (r) Sodium chloride 0.9%.	2. The medicine shall— (a) in the case of Lidocaine and Lidocaine hydrochloride, be administered only while attending on a woman in childbirth, and (b) where administration is— (i) by a registered midwife, be administered in the course of their professional practice; (ii) by a student midwife— (aa) be administered under the direct supervision of a registered midwife; and (bb) not include Diamorphine, Morphine or Pethidine hydrochloride.
3. Persons who are authorised as members of a group by a group authority granted under regulations 8(3) or 9(3) of the Misuse of Drugs Regulations 2001(**11**) or, regulations 8(3) or 9(3) of the Misuse of Drugs Regulations (Northern Ireland) 2002(**12**), to supply a controlled drug by way of administration only.	3. Prescription only medicines that are specified in the group authority.	3. The administration shall be subject to such conditions and in such circumstances and to such extent as may be specified in the group authority.
4. The owner or master of a ship which does not carry a doctor on board as part of the ship's complement.	4. All prescription only medicines that are for parenteral administration.	4. The administration shall be only so far as is necessary for the treatment of persons on the ship.
5. Persons operating an occupational health scheme.	5. Prescription only medicines that are for parenteral administration sold or supplied to the person operating an occupational health scheme in response to an order in writing signed by a doctor or a registered nurse.	5. The prescription only is administered in the course of an occupational health scheme, and the individual administering the medicine is— (a) a doctor, or (b) a registered nurse acting in accordance with the written instructions of a doctor as to the circumstances in which prescription only medicines of the description in question are to be used.

Persons exempted	Prescription only medicines to which the exemption applies	Conditions
6. The operator or commander of an aircraft.	6. Prescription only medicines f or parenteral administration which have been sold or supplied to the operator or commander of the aircraft in response to an order in writing signed by a doctor.	6. The administration shall be only so far as is necessary for the immediate treatment of sick or injured persons on the aircraft and shall be in accordance with the written instructions of the doctor as to the circumstances in which prescription only medicines of the description in question are to be used on the aircraft.
7. Persons employed as qualified first-aid personnel on off-shore installations.	7. All prescription only medicines that are for parenteral administration.	7. The administration shall be only so far as is necessary for the treatment of persons on the installation.
8. Persons who are registered paramedics.	8. The following prescription only medicines for parenteral administration— (a) Diazepam 5 mg per ml emulsion for injection, (b) Succinylated Modified Fluid Gelatin 4 per cent intravenous infusion, (c) medicines containing the substance Ergometrine Maleate 500 mcg per ml with Oxytocin 5 iu per ml, but no other active ingredient, (d) prescription only medicines containing one or more of the following substances, but no other active ingredient—	8. The administration shall be only for the immediate, necessary treatment of sick or injured persons and in the case of prescription only medicine containing Heparin Sodium shall be only for the purpose of cannula flushing.
	(i) Adrenaline Acid Tartrate, (ii) Adrenaline hydrochloride, (iii) Amiodarone, (iv) Anhydrous glucose, (v) Benzlypenicillin, (vi) Compound Sodium Lactate Intravenous Infusion (Hartmann's Solution), (vii) Ergometrine Maleate, (viii) Furosemide, (ix) Glucose, (x) Heparin Sodium, (xi) Lidocaine Hydrochloride, (xii) Metoclopramide, (xiii) Morphine Sulphate, (xiv) Nalbuphine Hydrochloride, (xv) Naloxone Hydrochloride, (xvi) Ondansetron (xvii) Paracetamol, (xviii) Reteplase, (xix) Sodium Chloride, (xx) Streptokinase, (xxi) Tenecteplase.	

Persons exempted	Prescription only medicines to which the exemption applies	Conditions
9. Persons who hold the advanced life support provider certificate issued by the Resuscitation Council (UK).	9. The following prescription only medicines for parenteral administration — (a) Adrenaline 1:10,000 up to l mg; and (b) Amiodarone.	9. The administration shall be only in an emergency involving cardiac arrest, and in the case of adrenaline the administration shall be intravenous only.
10. Persons ("P") who are members of Her Majesty's armed forces.	10. All prescription only medicines.	10. The administration shall be— (a) in the course of P undertaking any function as a member of Her Majesty's armed forces; and (b) where P is satisfied that it is not practicable for another person who is legally entitled to administer a prescription only medicine to do so; and (c) only in so far as is necessary— (i) for the treatment of a sick or injured person in an emergency, or (ii) to prevent ill-health where there is a risk that a person would suffer ill-health if the prescription only medicine is not administered.

PART 4 Exemptions from the restrictions in regulations 220 and 221 for certain persons who sell, supply, or offer for sale or supply certain medicinal products

Persons exempted	Medicinal products to which exemption applies	Conditions
1. Registered chiropodists and podiatrists.	1. Medicinal products on a general sale list which are for external use and are not veterinary drugs and the following pharmacy medicines for external use— (a) Potassium permanganate crystals or solution; (b) ointment of heparinoid and hyaluronidase; and (c) products containing, as their only active ingredients, any of the following substances, at a strength, in the case of each substance, not exceeding that specified in relation to that substance— (i) 9.0 per cent Borotannic complex (ii) 10.0 per cent Buclosamide (iii) 3.0 per cent Chlorquinaldol (iv) 1.0 per cent Clotrimazole (v) 10.0 per cent Crotamiton (vi) 5.0 per cent Diamthazole hydrochloride (vii) 1.0 per cent Econazole nitrate (viii) 1.0 per cent Fenticlor (ix) 10.0 per cent Glutaraldehyde	

Persons exempted	Medicinal products to which exemption applies	Conditions
	(x) 1.0 per cent Griseofulvin (xi) 0.4 per cent Hydrargaphen (xii) 2.0 per cent Mepyramine maleate (xiii) 2.0 per cent Miconazole nitrate (xiv) 2.0 per cent Phenoxypropan-2-ol (xv) 20.0 per cent Podophyllum resin (xvi) 10.0 per cent Polynoxylin (xvii) 70.0 per cent Pyrogallol (xviii) 70.0 per cent Salicylic acid (xix) 1.0 per cent Terbinafine (xx) 0.1 per cent Thiomersal.	
2. Registered chiropodists and podiatrists against whose names are recorded in the relevant register annotations signifying that they are qualified to use the medicines in column 2.	2. (a) The following prescription only medicines— (i) Amorolfine hydrochloride cream where the maximum strength of the Amorolfine in the cream does not exceed 0.25 per cent by weight in weight, (ii) Amorolfine hydrochloride lacquer where the maximum strength of Amorolfine in the lacquer does not exceed 5 per cent by weight in volume, (iii) Amoxicillin, (iv) Co-Codamol, (v) Co-dydramol 10/500 tablets, (vi) Codeine Phosphate, (vii) Erythromycin, (viii) Flucloxacillin, (ix) Silver Sulfadiazine, (x) Tioconazole 28%, (xi) Topical hydrocortisone where the maximum strength of the hydrocortisone in the medicinal product does not exceed 1 per cent by weight in weight; and (b) Ibuprofen, other than preparations of ibuprofen which are prescription only medicines.	2. The sale or supply shall be only in the course of their professional practice, and the medicinal product must have been made up for sale or supply in a container elsewhere than at the place at which it is sold or supplied.
3. Registered optometrists.	3. All medical products on a general sale list, all pharmacy medicines and prescription only medicines which are not for parenteral administration and which— (a) are eye drops and are prescription only medicines by reason only that they contain not more than— (i) 30.0 per cent Sulphacetamide Sodium, or (ii) 0.5 per cent Chloramphenicol, or	3. The sale or supply shall be only— (a) in the case of medicinal products on a general sale list and pharmacy medicines, in the course of their professional practice; (b) in the case of prescription only medicines, in the course of their professional practice and in an emergency.

Persons exempted	Medicinal products to which exemption applies	Conditions
	(b) are eye ointments and are prescription only medicines by reason only that they contain not more than— (i) 30.0 per cent Sulphacetamide Sodium, or (ii) 1.0 per cent Chloramphenicol, or (c) are prescription only medicines by reason only that they contain any of the following substances— (i) Cyclopentolate hydrochloride, (ii) Fusidic acid, (iii) Tropicamide.	
4. Additional supply optometrists.	4. Medicinal products which are prescription only medicines by reason only that they contain any of the following substances— (a) Acetylcysteine, (b) Atropine sulphate, (c) Azelastine hydrochloride, (d) Diclofenac sodium, (e) Emedastine, (f) Homotropinehydrobromide, (g) Ketotifen, (h) Levocabastine, (i) Lodoximide, (j) Nedocromil sodium, (k) Olopatadine, (l) Pilocarpine hydrochloride, (m) Pilocarpine nitrate, (n) Polymyxin B/bacitracin, (o) Polymyxin B/trimethoprim, (p) Sodium Cromoglycate.	4. The sale or supply shall be only in the course of their professional practice and only in an emergency.
5. Holders of manufacturer's licences where the licence in question contains a provision that the licence holder shall manufacture the medicinal product to which the licence relates only for a particular person after being requested by or on behalf of that person and in that person's presence to use his own judgement as to the treatment required.	5. Medicinal products on a general sale list which are for external use and are not veterinary drugs and pharmacy medicines which are for external use in the treatment of hair and scalp conditions and which contain any of the following— (a) not more than 5.0 per cent of Boric acid, (b) Isopropyl myristate or Lauryl sulphate, (c) not more than 0.004 per cent Oestrogens, (d) not more than 1.0 per cent of Resorcinol, (e) not more than 3.0 per cent of Salicylic acid, (f) not more than 0.2 per cent of Sodium pyrithione.	5. The licence holder shall sell or supply the medicinal product in question only to a particular person after being requested by or on behalf of that person and in that person's presence to use his own judgement as to the treatment required.

Persons exempted	Medicinal products to which exemption applies	Conditions
6. Persons selling or supplying medicinal products to universities, other institutions concerned with higher education or institutions concerned with research.	6. All medicinal products.	6. The sale or supply shall be— (a) Subject to the presentation of an order signed by the principal of the institution concerned with education or research or the appropriate head of department in charge of the specified course of research stating— (i) the name of the institution for which the medicinal product is required, (ii) the purpose for which the medicinal product is required, and (iii) the total quantity required, and (b) for the purposes of the education or research with which the institution is concerned.
7. Persons selling or supplying medicinal products to organisations for research purposes.	7. All medicinal products.	7. The sale or supply is only for the purposes of research and shall be— (a) subject to the presentation of an order signed by the representative of the organisation concerned stating— (i) who requires the medicine, (ii) the purposes for which it is required, (iii) the quantity required, and (iv) the purposes of the research with which the organisation is concerned; and (b) not for administration to humans.
8. Persons selling or supplying medicinal products to any of the following— (a) a public analyst appointed under section 27 of the Food Safety Act 1990 or under article 27 of the Food Safety (Northern Ireland) Order 1991; (b) an agricultural analyst appointed under section 67 of the Agriculture Act 1970(**13**),	8. All medicinal products.	8. The sale or supply is in connection with the exercise of any statutory function carried out by any person listed in sub-paragraphs (a) to (d) of column 1 provided that— (a) the medicinal products are requested on an order signed by or on behalf of a person listed in sub-paragraph (a) to (d) of column 1, and

Persons exempted	Medicinal products to which exemption applies	Conditions
(c) a person duly authorised by an enforcement authority under regulations 325 to 328, (d) a sampling officer within the meaning a sampling officer within the meaning of Schedule 31.		(b) the order gives— (i) the status of the person signing it, (ii) the amount of medicinal product required.
9. Holders of a marketing authorisation, a certificate of registration or a manufacturer's licence.	9. Medicinal product referred to in the marketing authorisation, certificate of registration or manufacturer's licence.	The sale or supply shall be only— (a) to a pharmacist, (b) so as to enable that pharmacist to prepare an entry relating to the medical product in question in a tablet or capsule identification guide or similar publication, and (c) of no greater quantity than is reasonably necessary for that purpose.
10. Registered dispensing opticians.	10. Pharmacy medicines for external use containing chloramphenicol at a strength not exceeding— (a) 0.5 per cent in eye drops; (b) 1 per cent in ointment.	10. The sale or supply shall only be in the course of their professional practice.
11. Operator or commander of an aircraft.	11. All medicinal products on a general sale list.	11. The medicinal product must— (a) have been made up for sale or supply in a container elsewhere than at the place at which it is sold or supplied; and (b) be stored in a part of the aircraft which the operator is able to close so as to exclude the public.
12. the operator of a train.	12. All medicinal products on a general sale list.	12. The medicinal product must— (a) have been made up for sale or supply in a container elsewhere than at the place at which it is sold or supplied; and (b) be stored in a part of the train which the operatoris able to close so as to exclude the public.
13. Registered orthoptists.	13. All medicinal products on a general sale list, all pharmacy medicines and the following prescription only medicines– (a) Atropine, (b) Cyclopentolate, (c) Tropicamide,	13. The sale or supply shall be only in the course of their professional practice.

Persons exempted	Medicinal products to which exemption applies	Conditions
	(d) Lidocaine with fluorescein, (e) Oxybuprocaine, (f) Proxymetacaine, (g) Tetracaine, (h) Chloramphenicol, (i) Fusidic acid.	

PART 5 Exemptions from the restrictions in regulations 220 and 221 for certain persons who supply certain medicinal products

Persons exempted	Medicinal products to which exemption applies	Conditions
1. Royal National Lifeboat Institution and certificated first aiders of the Institution.	1. All medicinal products.	1. The supply shall be only so far as is necessary for the treatment of sick or injured persons.
2. British Red Cross Society and certificated first aid and certificated nursing members of the Society.	2. All pharmacy medicines and all medicinal products on a general sale list.	2. The supply shall be only so far as is necessary for the treatment of sick or injured persons.
3. St John Ambulance Association and Brigade and certificated first aid and certificated nursing members of the Association and Brigade.	3. All pharmacy medicines and all medicinal products on a general sale list.	3. The supply shall be only so far as is necessary for the treatment of sick or injured persons.
4. St. Andrew's Ambulance Association and certificated first aid and certificated nursing members of the Association.	4. All pharmacy medicines and all medicinal products on a general sale list.	4. The supply shall be only so far as is necessary for the treatment of sick and injured persons.
5. Order of Malta Ambulance Corps and certificated first aid and certificated nursing members of the Corps.	5. All pharmacy medicines and all medicinal products on a general sale list.	5. The supply shall be only so far as is necessary for the treatment of sick or injured persons.
6. Persons authorised by licences granted under regulation 5 of the Misuse of Drugs Regulations 2001 or regulation 5 of the Misuse of Drugs Regulations (Northern Ireland) 2002.	6. Such prescription only medicines and such pharmacy medicines as are specified in the licence.	6. The supply shall be subject to such conditions and in such circumstances and to such an extent as may be specified in the licence.
7. Persons employed or engaged in the provision of lawful drug treatment services.	7. Ampoules of sterile water for injection that contain no more than 5ml of water each.	7. The supply shall be only in the course of provision of lawful drug treatment services.

Persons exempted	Medicinal products to which exemption applies	Conditions
8. Persons requiring medicinal products for the purpose of enabling them, in the course of any business carried on by them, to comply with any requirements made by or in pursuance of any enactment with respect to the medical treatment of their employees.	8. Such prescription only medicines and such pharmacy medicines as may be specified in the relevant enactment and medicinal products on a general sale list.	8. The supply shall be— (a) for the purpose of enabling compliance with any requirement made by or in pursuance of any such enactment, and (b) subject to such conditions and in such circumstances as may be specified in the relevant enactment.
9. The owner or master of a ship which does not carry a doctor on board as part of the ship's complement.	9. All medicinal products.	9. The supply shall be only so far as is necessary for the treatment of persons on the ship.
10. Persons operating an occupational health scheme.	10. All pharmacy medicines, all medicinal products on a general sale list and such prescription only medicines as are sold or supplied to a person operating an occupational health scheme in response to an order signed by a doctor or a registered nurse.	10. (a) The supply shall be in the course of an occupational health scheme. (b) The individual supplying the medicinal product, if not a doctor, shall be— (i) a registered nurse, and (ii) where the medicinal product in question is a prescription only medicine, acting in accordance with the written instructions of a doctor as to the circumstances in which prescription only medicines of the description in question are to be used in the course of an occupational health scheme.
11. Persons carrying on the business of a school providing full-time education.	11. Pharmacy medicines that are for use in the prevention of dental caries and consist of or contain Sodium Fluoride.	11. The supply shall be— (a) in the course of a school dental scheme, and (b) if to a child under 16 only where the parent or guardian of that child has consented to such supply.
12. Health authorities or Primary Health Trusts.	12. Pharmacy medicines that are for use in the prevention of dental caries and consist of or contain Sodium Fluoride.	12. The supply shall be in the course of— (a) a pre-school dental scheme, and the individual supplying the medicinal product shall be a registered nurse, or (b) a school dental scheme, and if to a child under 16 only where the parent or guardian of that child has consented to such supply.
13. The operator or commander of an aircraft.	13. All pharmacy medicines, all medicinal products on a general sale list and such prescription only medicines which are not for parenteral administration and which have been sold or supplied to the operator or commander of an aircraft in response to an order in writing signed by a doctor.	13. The supply shall be only so far as is necessary for the immediate treatment of sick or injured persons on the aircraft and, in the case of a prescription only medicine, shall be in accordance with the written instructions of a doctor as to the circumstances in which the prescription only medicines of the description in question are to be used on the aircraft.

Persons exempted	Medicinal products to which exemption applies	Conditions
14. Persons employed as qualified first-aid personnel on offshore installations.	14. All medicinal products.	14. The supply shall be only so far as is necessary for the treatment of persons on the installation.
15. A prison officer.	15. All medicinal products on the general sale list.	15. The supply shall only be so far as is necessary for the treatment of prisoners.
16. Persons who hold a certificate in first aid from the Mountain Rescue Council of England and Wales, or from the Northern Ireland Mountain Rescue Co-ordinating Committee.	16. All pharmacy medicines, all medicinal products on a general sale list and such prescription only medicines which are sold or supplied to a person specified in column 1 of this paragraph in response to an order in writing signed by a doctor.	16. The supply shall be only so far as is necessary for the treatment of sick or injured persons in the course of providing mountain rescue services.
17. Her Majesty's armed forces.	17. All medicinal products.	17. The supply shall be only so far as is necessary for the treatment of a sick or injured person or the prevention of ill-health.
18. A person ("P") carrying on the business of a school who is trained to administer the relevant medicine.	18. A prescription only medicinal product comprising an inhaler containing salbutamol.	18. The supply shall be— (a) in the course of P carrying on the business of a school; (b) where supply is to a pupil at that school who is known to suffer from asthma; and (c) where the pupil requires the medicinal product in an emergency.

(1) 1990 c.16. Section 27 was amended by the Local Government etc (Scotland) Act 1994 section 180(1) and Schedule 18 paragraph 163(3), the Food Standards Act 1999 section 40 (1) and Schedule 5 paragraphs 7 and 8, the Local Government (Wales) Act 1994 section 22 (3) and Schedule 9 paragraph 16(2), S.I. 1994/865 regulation 24, and the Local Government and Public Involvement in Health Act 2007 sections 22 and 241, Schedule 1 Part 2 paragraph 17, and Schedule 18 Part 1.
(2) 1991 No. 762 (N.I. 7). There are amendments not relevant to these Regulations.
(3) 1990 c.16.
(4) 1989 No. 846 (N.I. 6).
(5) 2006 c. 41.
(6) 1978 c. 29.
(7) 2006 c. 42.
(8) S.I. 1972/1265 (N.I. 14).
(9) S.I. 2001/3998, to which there are amendments that are not relevant.
(10) S.R. 2002 No. 1, to which there are amendments that are not relevant.
(11) S.I. 2001/3998 as amended by S.I. 2007/2154. There are other amendments that are not relevant.
(12) S.R. 2002 No. 1, as amended by S.R. 2007 No. 348. There are other amendments that are not relevant.
(13)1970 c.40: subsection (1) was amended by section 272(1) of and Schedule 30 to the Local Government Act 1972; section 16 of and Schedule 8 paragraph 15 to the Local Government Act 1985, and section 66(6) and (8) of, and Schedule 16 paragraph 38(5) and Schedule 18 to the Local Government (Wales) Act 1994. Subsection (1A) was inserted by section 66(6) of and Schedule 16 paragraph 38(5) to that Act. Subsection 2 was substituted by section 180(1) of and Schedule 13 paragraph 85(2) to the Local Government etc (Scotland) Act 1994, and subsection (7) was repealed by sections 1(1) and 194 of, and Schedule 1 paragraph 8 and Schedule 34 Part 1 to the Local Government, Planning and Land Act 1980.

Appendix 3

SOURCES OF USEFUL INFORMATION

The Medicines and Healthcare products Regulatory Agency regulates medicines, medical devices and blood components for transfusion in the UK:
https://www.gov.uk/government/organisations/medicines-and-healthcare-products-regulatory-agency

Register of authorised online sellers of medicines:
http://medicine-seller-register.mhra.gov.uk/

The Home Office plays a fundamental role in the security and economic prosperity of the United Kingdom:
https://www.gov.uk/government/organisations/home-office

The UK Border Force is a law enforcement command within the Home Office. It secures the UK border by carrying out immigration and customs controls for people and goods entering the UK:
https://www.gov.uk/government/organisations/border-force

In England the Care Quality Commission monitors, inspects and regulates health and social care services:
http://www.cqc.org.uk/

The Regulation and Quality Improvement Authority is responsible for registering and inspecting independent hospitals, clinics and other care services in Northern Ireland:
http://www.rqia.org.uk/home/index.cfm

Healthcare Improvement Scotland regulates independent specialist clinics and healthcare services in Scotland:
http://www.healthcareimprovementscotland.org

Healthcare Inspectorate Wales is the regulator of independent healthcare in Wales:
http://www.hiw.org.uk

The General Medical Council's register of doctors, including "Licence to practise" information:
http://www.gmc-uk.org/doctors/register/LRMP.asp

The General Pharmaceutical Council's register of pharmacy premises, pharmacists and pharmacy technicians:
http://www.pharmacyregulation.org/registers

The Pharmaceutical Society of Northern Ireland's register of pharmacy premises, pharmacists and pharmacy technicians:
http://www.psni.org.uk/search-register/

The Health & Care Professions Council register:
http://www.hpc-uk.org/aboutregistration/theregister/

The Royal College of Veterinary Surgeons' register:
http://www.rcvs.org.uk/registration/check-the-register/

The General Dental Council's register:
http://www.gdc-uk.org/Pages/SearchRegisters.aspx

Public Health England's "Green Book" relating to vaccines:
https://www.gov.uk/government/collections/immunisation-against-infectious-disease-the-green-book

UK legislation:
http://www.legislation.gov.uk/

European medicines legislation and guidelines:
http://ec.europa.eu/health/documents/eudralex/index_en.htm

European Medicines Agency:
http://www.ema.europa.eu/ema/

EudraGMDP database:
http://eudragmdp.ema.europa.eu/inspections/displayWelcome.do

World Health Organization guidelines – distribution:
http://www.who.int/medicines/areas/quality_safety/quality_assurance/distribution/en/

World Health Organization "List of globally identified websites of medicines regulatory authorities" (as of November 2012):
http://www.who.int/medicines/areas/quality_safety/regulation_legislation/list_mra_websites_nov2012.pdf?ua=1

Pharmaceutical Inspection Convention and Pharmaceutical Inspection Co-operation Scheme (PIC/S):
http://www.picscheme.org/

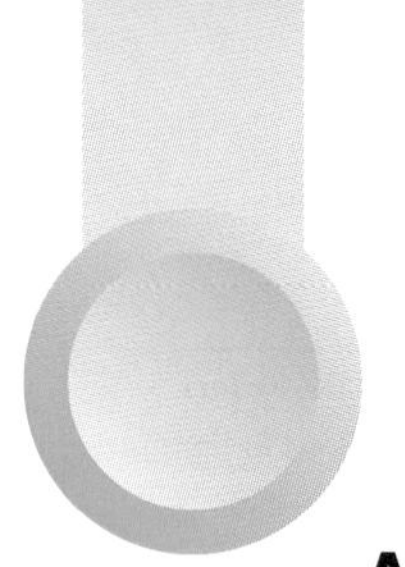

Appendix 4

LICENSING FOR IMPORT INTO THE UK AND EXPORT FROM THE UK INCLUDING INTRODUCED MEDICINE – WHOLESALE SUPPLY ONLY

I buy a medicine from a licensed manufacturer/ importer or distributor in the:	My destination market where the medicine is to be used is:	The medicine is:	The UK licence required to distribute the medicine is a:
UK	UK	Subject of a UK national MA	WDA(h)
		Subject of an EEA (non-UK) national MA	
		Centrally authorised[1]	
		Unlicensed 'specials' (Article 5(1) product)	
	EEA (non-UK)	Subject of a UK national MA	WDA(h)
		Subject of an EEA (non-UK) national MA	
		Centrally authorised[1]	
		Unlicensed 'specials' (Article 5(1) product)	
	3rd country	Subject of a UK national MA	WDA(h)
		Subject of an EEA (non-UK) national MA	
		Centrally authorised	
		Unlicensed - manufactured for export[2]	

[1]If not in livery of destination market, product will be subject to EMA parallel distribution scheme. Must be relabelled/repackaged by holder of MIA.
[2]Product must have been manufactured by an MIA holder for export. Unlicensed 'specials' (Article 5(1) products) cannot be exported as these have been manufactured by an MS holder.

I buy a medicine from a licensed manufacturer/ importer or distributor in the:	My destination market where the medicine is to be used is:	The medicine is:	The UK licence required to distribute the medicine is a:
EEA (non-UK)	UK	Subject of a UK national MA	WDA(h)
		Subject of an EEA (non-UK) national MA	
		Centrally authorised[3]	
		Unlicensed 'specials' (Article 5(1) product)	
	EEA (non-UK)	Subject of a UK national MA	WDA(h)
		Subject of an EEA (non-UK) national MA	
		Centrally authorised[3]	
		Unlicensed 'specials' (Article 5(1) product)	
	3rd country	Subject of a UK national MA	WDA(h)
		Subject of an EEA (non-UK) national MA	
		Centrally authorised	
		Unlicensed - manufactured for export[4]	
3rd country	UK	Subject of a UK national MA	MIA
		Subject of an EEA (non-UK) national MA	
		Centrally authorised	
		Unlicensed 'specials' (Article 5(1) product)	MS
	EEA (non-UK)	Subject of a UK national MA	MIA
		Subject of an EEA (non-UK) national MA	
		Centrally authorised	
		Unlicensed 'specials' (Article 5(1) product)	MS
	3rd country	Unlicensed[5]	WDA(h)
	3rd country (via another legal entity in the UK/EEA)	Unlicensed[6]	MIA

[3]If not in livery of destination market, product will be subject to EMA parallel distribution scheme. Must be relabelled/repackaged by holder of MIA.
[4]Product must have been manufactured by an MIA holder for export. Unlicensed 'specials' (Article 5(1) products) cannot be exported as these have been manufactured by an MS holder with no QP certification.
[5]Products have been 'introduced' into the UK without being officially imported and are subsequently exported to a 3rd country by the importer.
[6]Products sold or supplied to another entity in the UK/EEA who then goes on to export the product to a 3rd country.

Key:
WDA(h) – Wholesale Distribution Authorisation for human medicines
MIA – Manufacturer's/Importer's Authorisation
MS – Manufacturer (Specials) licence

EU Member States:
Austria
Belgium
Bulgaria
Croatia
Cyprus
Czech Republic
Denmark
Estonia
Finland
France
Germany
Greece
Hungary
Ireland
Italy
Latvia
Lithuania
Luxembourg
Malta
Netherlands
Poland
Portugal
Romania
Slovakia
Slovenia
Spain
Sweden
United Kingdom

EEA Member States:
All EU Member States and Iceland, Norway and Liechtenstein

UK Crown Dependencies:
Treated as part of the EEA for import/export:
Jersey

Guernsey
Isle of Man
Treated as 3rd country for import/export:
Gibraltar

UK Overseas Territories (all treated as third countries for import/export):
Anguilla
Bermuda
British Antarctic Territory
British Indian Ocean Territory
British Virgin Islands
Cayman Islands
Falkland Islands
Montserrat
Pitcairn Islands
Turks and Caicos Islands
Saint Helena, Ascension and Tristan da Cunha
South Georgia and the South Sandwich Islands

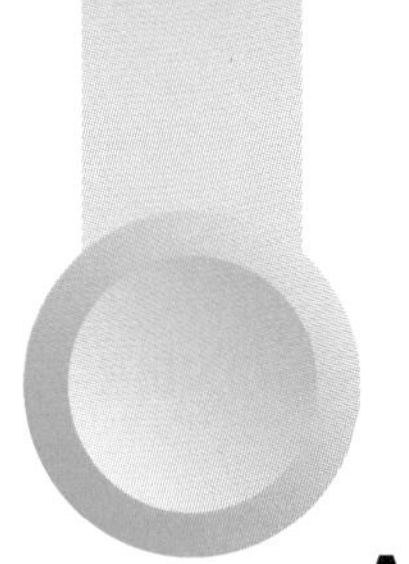

Appendix 5

IMPORTATION OF ACTIVE SUBSTANCES FOR MEDICINAL PRODUCTS FOR HUMAN USE – QUESTIONS AND ANSWERS

In June 2016 the Commission published a revised question and answer document responding to frequently asked questions in relation to the import of active substances. The revised document is produced here which includes the addition of Q&A 35 to clarify the requirements in case of importation of active substances released for sale before the expiration date of their written confirmation and minor editing of Q&A 10A and 29A.

The views expressed in this questions and answers document are not legally binding. Ultimately, only the European Court of Justice can give an authoritative interpretation of Union law.

This documents sets out frequently-asked 'questions and answers' regarding the new rules for the importation of active substances for medicinal products for human use.

These rules are contained in Articles 46b and 111b of Directive 2001/83/EC.

The 'written confirmation' is addressed in Article 46b(2)(b) of Directive 2001/83/EC.

1. QUESTION: WHEN DO THE NEW RULES FOR THE WRITTEN CONFIRMATION APPLY?

Answer: They apply as of 2 July 2013. Any active substance imported into the EU from that date is subject to the rules on the written confirmation.

2. QUESTION: DO THE RULES ON THE WRITTEN CONFIRMATION ALSO APPLY TO ACTIVE SUBSTANCES FOR VETERINARY MEDICINAL PRODUCTS?

Answer: No. The rules apply only to active substances for medicinal products for human use.

2A. QUESTION: DO THE RULES ON THE WRITTEN CONFIRMATION ALSO APPLY TO BLOOD PLASMA?

Answer: No. However, processed derivatives of plasma having a pharmacological, immunological or metabolic action are considered as active substance and written confirmation is thus required.

3. QUESTION: DO THE RULES ON THE WRITTEN CONFIRMATION APPLY TO ACTIVE SUBSTANCES FOR MEDICINAL PRODUCTS INTENDED FOR RESEARCH AND DEVELOPMENT TRIALS?

Answer: Active substances imported to be used in the manufacture of non-authorised medicinal products intended for research and development trials are excluded from the rules.

Active substances imported to be used in the manufacture of authorised medicinal products intended for research and development trials are expected to fulfil the requirements of Directive 2001/83/EC and be accompanied by a written confirmation, unless there is proof that the full amount of the imported API will be used for the manufacture of batches/units of an authorised medicinal product exclusively intended for research and development trials. In the latter case, those batches/units of an authorised medicinal product fall outside the scope of Directive 2001/83/EC and the API used in their manufacture is exempted from the rules on the written confirmation.

4. QUESTION: DO THE RULES ON THE WRITTEN CONFIRMATION APPLY TO ACTIVE SUBSTANCES WHICH ARE BROUGHT INTO THE EU WITHOUT BEING IMPORTED ('INTRODUCED' ACTIVE SUBSTANCES)? AN EXAMPLE IS THE INTRODUCTION OF AN ACTIVE SUBSTANCE WHICH IS SUBSEQUENTLY EXPORTED.

Answer: No. The rules on the written confirmation only apply to the import of active substances for medicinal products for human use.

5. QUESTION: WHAT IF, AT THE TIME OF EXPORT OF AN ACTIVE SUBSTANCE TO THE EU, IT IS NOT KNOWN WHETHER THE ACTIVE SUBSTANCE IS USED IN A MEDICINAL PRODUCT FOR HUMAN USE OR NOT?

Answer: If the consignment is not accompanied by a written confirmation, the active substance cannot be used in a medicinal product for human use.

6. QUESTION: IS THE WRITTEN CONFIRMATION EXPECTED TO CONFIRM COMPLIANCE WITH EU-RULES?

Answer: No. The written confirmation has to confirm compliance with GMP rules 'equivalent' to the rules applied in the EU.

7. QUESTION: IN MY NON-EU COUNTRY, THE APPLICABLE STANDARDS FOR MANUFACTURING OF ACTIVE SUBSTANCES ARE THE GOOD MANUFACTURING PRACTICES FOR ACTIVE SUBSTANCES OF THE WORLD HEALTH ORGANIZATION (WHO) - FORTY-FOURTH TECHNICAL REPORT, NO. 957, 2010, ANNEX 2. ARE THESE STANDARDS EQUIVALENT TO THOSE IN THE EU, AS REQUIRED ACCORDING TO EU LEGISLATION?

Answer: Yes.

8. QUESTION: IN MY NON-EU COUNTRY, THE APPLICABLE STANDARDS ARE ICH Q7. ARE THESE STANDARDS EQUIVALENT TO THOSE IN THE EU, AS REQUIRED ACCORDING TO EU LEGISLATION?

Answer: Yes.

9. QUESTION: DOES THE WRITTEN CONFIRMATION HAVE TO BE ISSUED BY A CENTRAL, REGIONAL OR LOCAL AUTHORITY?

Answer: Each non-EU country decides autonomously which body within that country issues the written confirmation. That non-EU country may decide to issue the written confirmation at central, regional or local level.

10. QUESTION: DO THE RULES APPLY ALSO TO ACTIVE SUBSTANCES CONTAINED IN AN IMPORTED FINISHED MEDICINAL PRODUCT?

Answer: No. Regarding finished medicinal products, the rules for importation of finished medicinal products (importation authorisation and batch release by a qualified person, see Articles 40(3) and 51 of Directive 2001/83/EC apply). These rules remain unchanged.

10A. QUESTION: IS WRITTEN CONFIRMATION ALSO REQUIRED FOR A STARTING MATERIAL OR AN INTERMEDIATE USED FOR THE PRODUCTION OF AN ACTIVE SUBSTANCE, FOR EXAMPLE BY WAY OF PURIFICATION OR FURTHER SYNTHESIS?

Answer: No. Such starting material or intermediate used for the production of an active substance does not fulfil the definition of Article 1(3a) of Directive 2001/83/EC.

11. QUESTION: IS THE WRITTEN CONFIRMATION ALSO REQUIRED FOR IMPORTED ACTIVE SUBSTANCES WHICH HAVE

ALREADY BEEN MIXED WITH EXCIPIENTS, WITHOUT YET BEING THE FINISHED MEDICINAL PRODUCT?

Answer: No. Such partial manufacturing of the finished product is not included in the rules on the written confirmation.

11A. QUESTION: IS THE WRITTEN CONFIRMATION ALSO REQUIRED WHERE THE FINISHED DOSAGE FORM MANUFACTURED IN THE EU IS DESTINED FOR EXPORTATION ONLY?

Answer: Yes.

12. QUESTION: WHO CHECKS THAT THE IMPORTED ACTIVE SUBSTANCE IS ACCOMPANIED BY THE WRITTEN CONFIRMATION?

Answer: This should be checked by the receiving manufacturer of the finished medicinal product. It may also be checked by the importer of the active substance upon its importation.

The verification whether such checks take place depends on the transposing law of the Member State where the active substance is imported. It may be verified:

– by the relevant authority upon importation; and/or
– in the context of an inspection of the importer of the active substance, and/or
– in the context of an inspection of the manufacturer of the medicinal product that uses the imported active substance.

13. QUESTION: HOW CAN I CHECK IF THE WRITTEN CONFIRMATION IS AUTHENTIC?

Answer: You should contact the manufacturer of the active substance or the issuing authority in the non-EU country.

14. QUESTION: IS THE WRITTEN CONFIRMATION SENT TO AN EU REGULATORY AGENCY?

Answer: No. The written confirmation accompanies the imported active substance.

15. QUESTION: DOES THE WRITTEN CONFIRMATION HAVE TO BE SUBMITTED WITH A REQUEST FOR AUTHORISATION OF A MARKETING AUTHORISATION OF A MEDICINAL PRODUCT?

Answer: No.

16. QUESTION: IS THE WRITTEN CONFIRMATION TO BE ISSUED FOR EACH BATCH/CONSIGNMENT'?

Answer: No. The written confirmation is issued per manufacturing plant and the active substance(s) manufactured on this site.

17. QUESTION: DOES EACH IMPORTED CONSIGNMENT HAVE TO BE ACCOMPANIED BY THE WRITTEN CONFIRMATION?

Answer: Yes.

18. QUESTION: IS IT ACCEPTABLE THAT THE WRITTEN CONFIRMATION ACCOMPANYING THE IMPORTED CONSIGNMENT OF THE ACTIVE SUBSTANCE IS A COPY?

Answer: Yes, provided that the original written confirmation is still valid.

18A. QUESTION: REGARDING THE WRITTEN CONFIRMATION OF 'EQUIVALENT' STANDARDS OF GOOD MANUFACTURING PRACTICE, CAN THE ISSUING AUTHORITY OF THE NON-EU COUNTRY BASE ITSELF ON INSPECTION RESULTS FROM EU AUTHORITIES OR OTHER AUTHORITIES APPLYING EQUIVALENT STANDARDS FOR GOOD MANUFACTURING PRACTICE, SUCH AS US FDA?

Answer: Yes. In this case, the written confirmation should indicate which authority has inspected the site.

18B. QUESTION: REGARDING THE WRITTEN CONFIRMATION OF 'EQUIVALENT' STANDARDS OF GOOD MANUFACTURING PRACTICE, CAN THE ISSUING AUTHORITY OF THE NON-EU COUNTRY BASE ITSELF ON INSPECTIONS CONDUCTED IN THE PAST?

Answer: Yes. It is not necessary to conduct an inspection specifically for the purpose of issuing the 'written confirmation'.

19. QUESTION:WHAT IS THE VALIDITY PERIOD OF THE WRITTEN CONFIRMATION?

Answer: The validity of the written confirmation is established by the issuing authority of the non-EU country.

19A. THE WRITTEN CONFIRMATION REFERS TO 'UNANNOUNCED INSPECTIONS'. DOES THIS MEAN THAT AN UNANNOUNCED INSPECTION HAS TO HAVE BEEN CONDUCTED?

Answer: No. Rather, the system of supervision as a whole (including different types of inspections, such as unannounced inspections) has to ensure a protection of public health at least equivalent to that in the EU.

20. QUESTION: IF ACTIVE SUBSTANCES ARE MANUFACTURED IN A NON-EU COUNTRY 'A', BUT IMPORTED IN THE EU VIA THE NON-EU COUNTRY 'B', WHO HAS TO ISSUE THE WRITTEN CONFIRMATION?

Answer: The written confirmation accompanying the imported active substance has to be issued by the non-EU country where the active substance is manufactured (i.e. non-EU country 'A').

21. QUESTION: THE TEMPLATE FOR THE WRITTEN CONFIRMATION REFERS TO A 'CONFIRMATION NUMBER'. DOES THIS NUMBER HAVE TO BE A SEQUENTIAL NUMBER PER COUNTRY?

Answer: No. This number would be attributed by the issuing authority of the non-EU country.

22. QUESTION: THE TEMPLATE FOR THE WRITTEN CONFIRMATION REFERS TO A 'RESPONSIBLE PERSON' IN THE ISSUING AUTHORITY. DOES THIS RESPONSIBLE PERSON HAVE TO HAVE A SPECIFIC QUALIFICATION?

Answer: No. The 'responsible person' in this context is the person responsible within the administration for issuing the written confirmation.

23. QUESTION: ACCORDING TO THE TEMPLATE FOR THE WRITTEN CONFIRMATION, INFORMATION OF FINDINGS RELATING TO NON-COMPLIANCE ARE SUPPLIED TO THE EU. TO WHOM THIS INFORMATION SHOULD BE SENT TO?

Answer: The information should be sent to the European Medicines Agency (qdefect@ema.europa.eu).

24. QUESTION: IS THE WRITTEN CONFIRMATION ALSO REQUIRED WHERE THERE IS A 'MUTUAL RECOGNITION AGREEMENT' BETWEEN A NON-EU COUNTRY AND THE EU?

Answer: Yes. The process of a written confirmation is independent of the existence of 'mutual recognition agreements'.

25. QUESTION: IF A MANUFACTURING PLANT IS LOCATED IN A NON-EU COUNTRY 'A', CAN THE WRITTEN CONFIRMATION BE ISSUED BY AN AUTHORITY IN ANOTHER NON-EU COUNTRY (NON-EU COUNTRY 'B')?

Answer: No.

26. QUESTION: ARE THERE EXCEPTIONS FROM THE REQUIREMENT OF A WRITTEN CONFIRMATION?

Answer: The Commission publishes a list of countries which, following their request, have been assessed and are considered as having equivalent rules for good manufacturing practices to those in the EU. Active substances manufactured in these countries do not require a written confirmation.

See also Questions 27 and 28.

27. QUESTION: WHERE CAN I FIND THE LIST OF NON-EU COUNTRIES TO WHICH THE REQUIREMENT OF A WRITTEN CONFIRMATION DOES NOT APPLY?

Answer: The list is published in the *Official Journal of the European Union* and also reproduced here: http://ec.europa.eu/health/human-use/quality/index_en.htm.

28. QUESTION: HOW MANY NON-EU COUNTRIES HAVE SO FAR REQUESTED TO BE LISTED?

Answer: A list of non-EU countries which have so far requested to be listed is available here: http://ec.europa.eu/health/human-use/quality/index_en.htm.

29. QUESTION: WHEN IS THE LIST GOING TO BE PUBLISHED BY THE COMMISSION?

Answer: The Commission is going to publish an additional non-EU country on the list once its equivalence assessment has been finalised. The equivalence-assessment takes several months from the request from the non-EU country.

29A. QUESTION: HOW DOES A NON-EU COUNTRY REQUEST TO BE LISTED?

Answer: The request is made by way of a letter to the Director-General of DG SANTE. It should contain the relevant information for conducting the 'equivalence assessment'.

More information on the procedure and the documents to be submitted is available here: http://ec.europa.eu/health/human-use/quality/index_en.htm, under the section "Listing of third countries".

The relevant information can also be sent directly to the responsible service within the Commission (sante-pharmaceuticals-b4@ec.europa.eu).

30. QUESTION: DO I NEED A WRITTEN CONFIRMATION, EVEN THOUGH MY MANUFACTURING SITE HAS RECENTLY BEEN INSPECTED BY THE EUROPEAN DIRECTORATE FOR THE QUALITY OF MEDICINES (EDQM) OF THE COUNCIL OF EUROPE?

Answer: Yes. The process of a written confirmation is independent of such inspection activities. See also Question 31.

31. QUESTION: DO I NEED A WRITTEN CONFIRMATION, EVEN THOUGH MY MANUFACTURING SITE HAS RECENTLY BEEN INSPECTED BY AN EU MEMBER STATE?

Answer: Yes. The process of a written confirmation is independent of such inspection activities. However, exceptionally and where necessary to ensure the availability of medicinal products, following inspections by an EU Member State, a Member State may decide to waive the need for a written confirmation for a period not exceeding the validity of the GMP certificate ('waiver').

32. QUESTION: I WOULD LIKE TO BE INSPECTED BY AN EU MEMBER STATE. WHERE DO I 'APPLY' FOR SUCH AN INSPECTION?

Answer: You should address through

- any registered importer of the active substance;
- any holder of a manufacturing authorisation that uses the active substance;
- any holder of a marketing authorisation that lists the active substance manufacturer to the national competent authority of the EU Member State where they are established.

33. QUESTION: WHAT HAPPENS WHEN AN ACTIVE SUBSTANCE MANUFACTURING SITE COVERED BY A WRITTEN CONFIRMATION IS FOUND GMP NON-COMPLIANT FOLLOWING AN INSPECTION BY AN EU MEMBER STATE?

Answer: A statement of GMP non-compliance issued by a EU Member State for a specific site and API supersedes the corresponding written confirmation until the noncompliance is resolved.

34. QUESTION: WHERE CAN I FIND A LIST OF ACTIVE SUBSTANCE MANUFACTURING SITES THAT RECEIVED STATEMENTS OF GMP NON-COMPLIANCE?

Answer: Statements of GMP non-compliance are stored in the EudraGMDP database (http://eudragmdp.eudra.org/inspections/displayWelcome.do) and publicly available.

35. QUESTION: CAN AN API BATCH MANUFACTURED DURING THE PERIOD OF VALIDITY OF A WRITTEN CONFIRMATION BE IMPORTED INTO THE EU ONCE THE WRITTEN CONFIRMATION IS EXPIRED?

Answer: Article 46(b)(2)(b) sets out that active substances can only be imported if manufactured in accordance with EU GMP or equivalent, and accompanied by a written confirmation from the competent authority of the exporting third country certifying, inter alia, that (1) the GMP standards applicable to the manufacturing plant are equivalent to those of the EU, and (2) the supervision of the plant compliance with GMP ensures a protection of public health equivalent to that of the EU.

It is legitimate to consider that the guarantees of equivalence provided by the written confirmation apply to any API batch in the scope of the written confirmation which was released for sale within the period of validity of the written confirmation, even if not exported in that time period.

Against this background, it can therefore be considered that the importation into the EU of an API accompanied by an expired WC is acceptable provided that the paperwork accompanying the consignment (1) unequivocally proves that the whole consignment has been manufactured and released for sale by the quality unit before the expiry date of the written confirmation; and (2) provides a solid justification of why a valid written confirmation is not available.

Appendix 6

SAFETY FEATURES FOR MEDICINAL PRODUCTS FOR HUMAN USE – QUESTIONS AND ANSWERS

Contents

1. General 303
2. Technical Specifications of the Unique Identifier 306
3. General Provision on the Verification and Decommissioning of the Safety Features 309
4. Verification of the Safety Features and Decommissioning of the Unique Identifier by Manufacturers 310
5. Verification of the Safety Features and Decommissioning of the Unique Identifier by Wholesalers 310
6. Verification of the Safety Features and Decommissioning of the Unique Identifier by Persons Authorised or Entitled to Supply Medicinal Products to the Public 311
7. Establishment, Management and Accessibility of the Repositories System 312
8. Obligations of Marketing Authorisation Holders, Parallel Importers and Parallel Distributors 315
9. Lists of Derogations and Notifications to the Commission 316

In June 2016 the Commission published a revised question and answer document responding to frequently asked questions in relation to safety features for medicinal products.

The views expressed in this questions and answers document are not legally binding. Ultimately, only the European Court of Justice can give an authoritative interpretation of Union law.

This document sets out frequently-asked 'questions and answers' regarding the implementation of the rules on the safety features for medicinal products for human use.

These rules are enshrined in Articles 47a, 54(o) and 54a of Directive 2001/83/EC, and in the Commission Delegated Regulation (EU) No 2016/1612.

1. GENERAL

1.1. Question: What are the safety features?

Answer: The safety features consist of two elements placed on the packaging of a medicinal product:

(1) a unique identifier, a unique sequence carried by a two-dimensional barcode allowing the identification and authentication of the individual pack on which it is printed; and
(2) a device allowing the verification of whether the packaging of the medicinal product has been tampered with (anti-tampering device).

1.2. Question: When do the rules on the safety features apply?

Answer: They apply as of 9th February 2019. Belgium, Greece and Italy have the option of deferring the application of the rules by an additional period of up to 6 years.

1.3. Question: Do the safety features need to be applied on all medicinal products for human use?

Answer: No. The safety features should only be applied on the packaging of the following medicinal products for human use:

(1) medicinal products subject to prescription which are not included in the list set out in Annex I to of Regulation (EU) No 2016/161;
(2) medicinal products not subject to prescription included in the list set out in Annex II of Regulation (EU) No 2016/161.
(3) medicinal products to which Member States have extended the scope of the unique identifier or the anti-tampering device in accordance with Article 54a(5) of Directive 2001/83/EC.

1.4. Question: Are there exceptions from the requirements for certain medicinal products to bear or not the safety features?

Answer: Yes. The list of categories of medicinal products subject to prescription which shall not bear the safety features are set out in Annex I of Regulation (EU) No 2016/161, while the list of medicinal products not subject to prescription which shall bear the safety features are set out in Annex II of the same Regulation.

1.5. Question: Do the rules on the safety features also apply to veterinary medicinal products?

Answer: No. The rules apply only to medicinal products for human use.

1.6. Question: Do the rules on the safety features apply to medicinal products intended for research and development trials?

Answer: Medicinal products intended for research and development trials and not yet granted a marketing authorisation are excluded from the rules on the safety features.

Authorised medicinal products have to fulfil the requirements of Directive 2001/83/EC and Regulation (EU) No 2016/161 up to the moment it becomes known which batch/unit will be used for research and development trials. In practice, a batch of an authorised investigational medicinal product or an authorised auxiliary medicinal product is excluded from the rules on the safety features if it is known at the time of manufacture that the whole batch is manufactured for use in clinical trials.

In addition, unique identifiers on authorised investigational medicinal products and authorised auxiliary medicinal products bearing the safety features should be decommissioned in accordance with Articles 16 and 25 (4)(c) of Regulation (EU) No 2016/161.

1.7. Question: Are the safety features required where the medicinal product manufactured in the EU is destined for exportation only?

Answer: No.

1.8. Question: in the case of a medicinal product bearing the safety features is brought into the territory of a Member State in accordance with Article 5(1) of Directive 2001/83/EC, do the rules on the safety features apply?

Answer: When a medicinal product is brought into the territory of a Member State in accordance with Article 5(1) of Directive 2001/83/EC, the rules on the safety features do not apply.

In practical terms, when an medicinal product is brought into the territory of a Member State in accordance with Article 5(1), the "importer" of that product is not required to (re)place safety features on its packaging (e.g. through labelling/relabelling operations).

However, in case the medicinal product brought into the territory of a Member State in accordance with Article 5(1) already bears the safety features, pharmacies, healthcare institutions and other relevant stakeholders in that Member State are strongly encouraged to verify the authenticity of and decommission the medicinal product before supplying it to the public.

1.9. Question: Does an obligation to bear "the safety features" imply an obligation to bear both a unique identifier and an anti-tampering device?

Answer: Yes.

1.10. Question: Once Regulation (EU) No 2016/161 applies, can manufacturers place the safety features, on a voluntary basis, on medicinal products not required to bear the safety features?

Answer: No. Once Regulation (EU) No 2016/161 applies, manufacturers cannot place the safety features on medicinal products not required to bear the safety features, unless the Member States have extended the scope of application of the unique identifier or of the anti-tampering device to those medicinal products in accordance with Article 54a(5) of Directive 2001/83/EC.

1.11. Question: Certain medicinal products are currently bearing an anti-tampering device on a voluntary basis. Are those products allowed to maintain the anti-tampering device once Regulation (EU) No 2016/161 applies, if they are not required to bear the safety features?

Answer: Once Regulation (EU) No 2016/161 applies, medicinal products can only bear an anti-tampering device if they are in the scope of Article 54a(1) of Directive 2001/83/EC (i.e. if they are medicinal products subject to prescription or medicinal products listed in Annex II of Regulation (EU) No 2016/161) or if the Member State(s) where they are placed on the market extended the scope of the anti-tampering device to those medicinal products.

1.12. Question: Would it be possible to place a unique identifier on the packaging of a medicinal product during the 3 years period between the publication of Regulation (EU) No 2016/161 and its application?

Answer: Yes, on a voluntary basis. It is recommended that, whenever possible, unique identifiers are placed on the packaging only once a functional national/supranational repository allowing the storage, verification of the authenticity and decommissioning of those identifiers is in place. Unique identifiers which are placed on medicinal products before such repository is in place are expected to be uploaded in the repository as soon as it becomes operational.

1.13. Question: Will the mandatory changes to the packaging due to the placing of the unique identifier and of the anti-tampering device require the submission of variations to marketing authorisations?

Answer: The regulatory requirements to be followed to notify the EMA of the placing of the unique identifier and/or the anti-tampering device on centrally authorised products are detailed in an implementation plan developed by the EMA and the European Commission and published in the "product information templates" section of the EMA website: http://www.ema.europa.eu/docs/en_GB/document_library/Other/2016/02/WC500201413.pdf

The regulatory requirements for nationally authorised products are available on the HMA/CMDh website:

http://www.hma.eu/fileadmin/dateien/Human_Medicines/CMD_h_/Falsified_Medicines/CMDh_345_2016_Rev00_02_2016_1.pdf

1.14. Question: Are there any mandatory specifications for the anti-tampering device?

Answer: In accordance with Article 54(o) of Directive 2001/83/EC and Article 3(2)(2) of Regulation (EU) No 2016/161, an anti-tampering device has to allow the verification of whether the packaging of the medicinal product has been tampered with.

There are no other mandatory specifications. The CEN standard EN 16679:2014 "Tamper verification features for medicinal product packaging" is available for manufacturers to consider.

2. TECHNICAL SPECIFICATIONS OF THE UNIQUE IDENTIFIER

2.1. Question: Does Regulation (EU) No 2016/161 limit the length of the unique identifier to 50 characters?

Answer: No. Only the length of the product code, one of the data elements of the unique identifier, is limited to 50 characters.

2.2. Question: Would it be possible to include, on a voluntary basis, a two-dimensional barcode on the packaging of medicinal products for human use not having to bear the safety features if the information carried by the barcode does not serve the purposes of identification and authentication of the medicinal product and does not include a unique identifier?

Answer: Yes, provided that the relevant labelling provisions of Title V of Directive 20014/83/EC are complied with.

Examples may include two-dimensional barcodes encoding price indications, reimbursement conditions, etc.

2.3. Question: Is it possible to include one-dimensional barcodes on the packaging of medicinal products for human use having to bear the safety features, in addition to the two-dimensional barcode carrying the unique identifier?

Answer: Yes, provided that the inclusion of both barcodes does not negatively impact the legibility of the outer packaging.

2.4. Question: Is a printing quality of 1.5 according to ISO/IEC 15415 mandatory?

Answer: No. Manufacturers are required to use a printing quality which ensures the accurate readability of the Data Matrix throughout the supply chain until at least one year after the expiry date of the pack or five years after the pack has been released for sale or distribution in accordance with Article 51(3) of Directive 2001/83/EC, whichever is the longer period.

The use of a printing quality of 1.5 or higher gives a presumption of conformity, i.e. manufactures using a printing quality of 1.5 or higher will be presumed to have fulfilled the requirement mentioned in the first paragraph without need to prove that it is actually the case.

If a printing quality lower than 1.5 is used, manufacturers may be asked to prove that requirements mentioned in the first paragraph are met.

2.5. Question: Can manufacturers, on a voluntary basis, place the human readable code on medicinal products with packaging having the sum of the two longest dimensions equal or less than 10 centimetres?

Answer: Yes.

2.6. Question: Are medicinal products with packaging having the sum of the two longest dimensions equal or less than 10 centimetres exempted from bearing the two-dimensional barcode carrying the unique identifier?

Answer: No, Article 7(2) only provides for an exemption from bearing the unique identifier in human readable format. The unique identifier in machine-readable format – the 2D barcode – is still required.

2.7. Question: Is it compulsory to print the national reimbursement number in human-readable format?

Answer: The national reimbursement number or other national number should be printed in human readable format only if required by the national competent authorities of the relevant Member State and not printed elsewhere on the packaging. It should be printed adjacent to the two-dimensional barcode if the dimensions of the packaging allow it.

2.8. Question: Is it compulsory for the human-readable data elements of the unique identifier to be placed adjacent to the two-dimensional barcode?

Answer: Yes, whenever the dimensions of the packaging allow it.

2.9. Question: What is the smallest font size that can be used to print the unique identifier in human-readable format?

Answer: The font size of the unique identifier should be in accordance with the "Guideline on the readability of the labelling and package leaflet of medicinal products for human use" published in Eudralex – Notice to

Applicants – Volume 2C (http://ec.europa.eu/health/files/eudralex/vol-2/c/2009_01_12_readability_guideline_final_en.pdf).

2.10. Question: Should the data elements of the unique identifier in human-readable format be printed on the packaging (product code, serial number and, where applicable, national reimbursement number) in an established order?

Answer: Yes. The order of printing is set out in the QRD template and is the same as set out in Regulation (EU) No 2016/161: Product code, Serial number, National reimbursement number or other national number identifying the medicinal product (where required).

2.11. Question: Regulation (EU) No 2016/161 does not mention batch number and expiry date as mandatory components of the human readable code. Is it mandatory to print the batch number and the expiry date in a human-readable format and adjacent to the two dimensional barcode?

Answer: Batch number and expiry date are mandatory components of the labelling of all medicinal products – regardless of whether they bear the safety features – and should be printed on the packaging in accordance with Article 54 (h) and (m) of Directive 2001/83/EC. There is no obligation to place batch number and expiry date adjacent to the two dimensional barcode.

2.12. Question: Is it allowed to place a QR code on the packaging of a medicinal product bearing the safety features?

Answer: Regulation (EU) No 2016/161 does not prohibit the placing of a QR code as far as it is not used for the purposes of identification and authentication of medicinal products.

Marketing authorisation holders are however encouraged, wherever technically feasible, to exploit the residual storage capacity of the Data Matrix to include the information they would otherwise include in the QR code. This would minimise the number of visible barcodes on the packaging and reduce the risk of confusion as regard the barcode to be scanned for verifying the authenticity of the medicinal product.

2.13. Question: Where on the packaging should the unique identifier be placed?

Answer: The delegated Regulation does not specify where on the outer packaging the safety features should be placed. The placement of the safety features is therefore to be supervised by competent authorities in accordance with current practice for labelling requirements.

3. GENERAL PROVISION ON THE VERIFICATION AND DECOMMISSIONING OF THE SAFETY FEATURES

3.1. Question: How should the unique identifier be decommissioned if the two-dimensional barcode is unreadable or deteriorated?

Answer: The unique identifier in human readable format should be recorded by any suitable method allowing the subsequent manual querying of the repository system in order to verify and decommission the unique identifier.

3.2. Question: Where the barcode carrying the unique identifier cannot be read, or in case the verification of the unique identifier is temporarily impeded, is it possible to supply the medicinal product to the public?

Answer: Article 30 of Regulation (EU) No 2016/161 prohibits supply to the public if there is reason to believe that the packaging of the medicinal product has been tampered with, or the verification of the safety features of the medicinal product indicates that the product may not be authentic.

In all other cases, the supply of medicinal products to the public is regulated by national legislation.

Without prejudice to national legislation, in the case where it is permanently impossible to read the unique identifier and verify the authenticity of the medicinal product, for example because both the data matrix and the human readable code are damaged, it is recommended that the medicinal product is not supplied to the public.

3.3. Question: Can a medicinal product which cannot be authenticated be returned, and to whom? Who should pay for the return?

Answer: Regulation (EU) No 2016/161 does not change the national provisions in place regulating returns of medicines from pharmacies and hospitals. The regulation of returns of medicinal products, including their financial aspects, remains a national competence.

3.4. Question: Is it allowed to simultaneously verify the authenticity of or decommission multiple unique identifiers by scanning an aggregated code?

Answer: Yes, it is possible to verify the authenticity of or decommission multiple unique identifiers by scanning an aggregated code rather than scanning each individual pack, provided that the requirements of Regulation (EU) No 2016/161 are complied with.

4. VERIFICATION OF THE SAFETY FEATURES AND DECOMMISSIONING OF THE UNIQUE IDENTIFIER BY MANUFACTURERS

4.1. Question: Do the records referred to in Article 15 of Regulation (EU) No 2016/161 have to be stored in the repositories system?

Answer: No. The manufacturers can decide how and where to keep the records of every operation he performs with or on the unique identifier.

4.2. Question: Article 18 requires that, in case of suspected falsification or tampering, the manufacturer should inform the competent authorities. Should he also inform the holder of the marketing authorization for the medicinal product?

Answer: Yes, Article 46 of Directive 2001/83/EC requires manufacturers to inform the competent authority and the marketing authorisation holder immediately if they obtain information that medicinal products which come under the scope of their manufacturing authorisation are, or are suspected of being, falsified.

4.3. Question: Articles 18, 24 and 30 of Regulation (EU) No 2016/161 require that manufacturers, wholesalers and persons authorised or entitled to supply medicinal products to the public immediately inform national competent authorities in case of suspected falsification of medicinal products. How should this information be notified?

Answer: The delegated Regulation does not specify how this information should be notified to competent authorities. However, Article 117a of Directive 2001/83/EC requires that Member States have in place a system aiming at preventing medicinal products that are suspected to present a danger to health from reaching the patient, including suspected falsified medicinal products. Manufacturers, wholesalers and persons authorised or entitled to supply medicinal products to the public may therefore inform authorities by means of such system.

5. VERIFICATION OF THE SAFETY FEATURES AND DECOMMISSIONING OF THE UNIQUE IDENTIFIER BY WHOLESALERS

5.1. Question: How should the expression “the same legal entity” referred to in Articles 21(b) and 26(3) of Regulation (EU) No 2016/161 be interpreted?

Answer: This expression should be interpreted in accordance with national legislation. As general guidance, and without prejudice to national legislation, a legal entity may be considered the same when, for example, it has the same registration number in the national company registry or, if no national registration is required, the same number for tax purposes (i.e. VAT number).

5.2. Question: Member States may hold stocks of certain medicinal products for the purpose of public health protection. How should the unique identifiers on those products be verified and decommissioned?

Answer: In accordance with Article 23(f) of the delegated Regulation No 2016/161, Member States may request wholesalers to verify the safety features of and decommission the unique identifier of medicinal products which are supplied to governmental institutions maintaining stocks of medicinal products for the purposes of civil protection and disaster control.

5.3. Question: Articles 18, 24 and 30 of Regulation (EU) No 2016/161 require that manufacturers, wholesalers and persons authorised or entitled to supply medicinal products to the public immediately inform national competent authorities in case of suspected falsification of medicinal products. How should this information be notified?

Answer: It is recommended that manufacturers, wholesalers and persons authorised or entitled to supply medicinal products to the public contact national competent authorities to be informed about the correct procedure to follow for the notification, since such notification is a national competence.

6. VERIFICATION OF THE SAFETY FEATURES AND DECOMMISSIONING OF THE UNIQUE IDENTIFIER BY PERSONS AUTHORISED OR ENTITLED TO SUPPLY MEDICINAL PRODUCTS TO THE PUBLIC

6.1. Question: In-patients in a hospital may be administered medicinal products during their stay, the costs of which may be charged to their insurer, which constitutes a sale. In this case, would the hospital (or any other healthcare institution) be allowed to verify the safety features and decommission the unique identifier of those products earlier than the time of supply to the public, in accordance with Article 25(2)?

Answer: Yes. In the case described, the charging of the medicinal products costs to the patient's insurer happens as a consequence of the administration of that product to the patient (regardless of whether the sale takes place before or after the actual administration). Consequently, it is considered that the charging of the cost of the medicinal product to the patient's insurer (or to the patient himself, for the matter) does not preclude hospitals from applying the derogation provided for in Article 25(2).

6.2. Question: How should the expression "the same legal entity" referred to in Articles 21(b) and 26(3) of Regulation (EU) No 2016/161 be interpreted?

Answer: See Q&A 5.1.

6.3. Question: Many hospitals and other healthcare institutions supply the contents of packages of a medicinal product to more than one patient. Where only part of a pack of a medicinal product is supplied, when should the decommissioning of the unique identifier be performed?

Answer: The unique identifier should be decommissioned when the packaging is opened for the first time, as required by Article 28 of Regulation (EU) No 2016/161.

6.4. Question: Does automated dose dispensing require the placing of new safety features on the individual patient doses/packs?

Answer: No. Automated dose dispensing falls in the scope of Article 28 of Regulation (EU) No 2016/161. Consequently, it is not necessary to place new safety features on the individual patient's dose/pack.

7. ESTABLISHMENT, MANAGEMENT AND ACCESSIBILITY OF THE REPOSITORIES SYSTEM

7.1. Question: How should the expression "manufacturers of medicinal products bearing the safety features", as used in Regulation (EU) No 2016/161, be interpreted?

Answer: For the purposes of Regulation (EU) No 2016/161, "manufacturer" means the holder of a manufacturing authorisation in accordance with Article 40 of Directive 2001/83/EC. The expression "manufacturers of medicinal products bearing the safety features" encompasses any holder of the said authorisation performing partial or total manufacture of a medicinal product bearing the safety features.

7.2. Question: Article 31 of Regulation (EU) No 2016/161 allows wholesalers and persons authorised or entitled to supply medicinal products to the public to participate in the legal entity/ies setting up and managing the repositories system, at no costs. Can the terms of such participation be regulated by stakeholders, for example through the statutes of establishment or incorporation of the legal entity/ies?

Answer: Yes, it is possible, provided that the terms do not contradict what is enshrined in legislation. In case of discrepancy, the provisions of Regulation (EU) No 2016/161 and Directive 2001/83/EC prevail.

7.3. Question: What is a supranational repository?

Answer: In practice, a repository serving as "national" repository for more than one Member State.

7.4. Question: How should the expressions "application programming interface" or "graphical user interface" referred to in Articles 32(4) and 35(1) of Regulation (EU) No 2016/161 be interpreted?

Answer: The expression "application programming interface" refers to a software/software interface consisting of a set of programming instructions and standards used by a piece of software to ask another piece of software to perform a task. The programming instructions and standards are set by the software being called upon. In the context of Regulation (EU) No 2016/161, the expression refers to the programming instructions and standards allowing the software of persons authorised or entitled to supply medicines to the public, wholesalers and national competent authorities to query the repository system.

The expression "graphical user interface" (GUI) refers to a human/computer interface that allows users to interact with software or a database through graphical icons and visual indicators without the need of using complex programming language.

Article 35(1)(i) limits the use of the GUI by wholesalers and persons authorised/entitled to supply medicines to the public to the very specific case of failure of their own software. In practice, it can be considered that:

- wholesalers and persons authorised or entitled to supply medicines to the public are expected to use their own software to connect to their national repository and verify the authenticity of/decommission the unique identifier; and
- they should use the GUI for the above purposes exclusively when their software fails.

7.5. Question: Article 33(1), second subparagraph, requires that information referred to in paragraphs 2(a) to 2(d) of that article, with the exception of the serial number, is stored in the hub. Does this mean that the serial number cannot be uploaded to the hub?

Answer: No, the provision only regulates which information is to be stored in the hub.

7.6. Question: Articles 34(4), 35(4) and 36(n) refer to the linking of the information on unique identifiers removed or covered to the information on the equivalent unique identifiers placed for the purposes of complying with Article 47a of Directive 2001/83/EC. Is the linking required to be at the level of individual unique identifiers? How does the linking work in practice?

Answer: No, it is not necessary to link individual unique identifiers. The link can be made at batch level by linking the list of decommissioned unique identifiers in the "old" batch (the batch to be repacked/relabelled) and the list of new unique identifiers placed on packs in the "new" batch (the repacked batch). The provision does not require the linking to be done at the level of individual unique identifiers, since the number of packs in the batch to be repacked/relabelled (and consequently the number of unique identifiers in that batch) may not correspond to the number of packs (and of unique identifiers) in the new batch – making a one-to-one link between unique identifiers impossible.

7.7. Question: In Article 35(1)(f), does the upper limit of 300 ms for a repository to respond to queries also apply when multiple repositories are implicated in the query, for example in case of cross-border verification?

Answer: 300 ms is the maximum response time of an individual repository. When the verification/decommissioning operation requires the querying of multiple repositories in the repositories system, for example in case of cross-border verification, the maximum response time is obtained by multiplying the maximum response time of an individual repository (300 ms) by the number of repositories involved in the query – for example, the maximum response time for a query involving national repository A, the hub, and national repository B would be 900 ms.

It should be noted that the system response time does not include the time needed by the query data to move from one repository to the other (which depends on the speed of the internet connection).

7.8. Question: How will the identity, role and legitimacy of the users of the repository system be verified?

Answer: It is the responsibility of the legal entity establishing and managing a repository to put in place appropriate security procedures ensuring that only verified users, i.e. users whose identity, role and legitimacy has been verified, are granted access to that repository.

7.9. Question: In Article 38(1), does the sentence "with the exception of the information referred to in Article 33(2)" refer to data access only, or also to data ownership?

Answer: It refers to data access only.

7.10. Question: In Article 38(1), what is the meaning of "information on the status of the unique identifier"?

Answer: The information on the status of the unique identifier includes whether the unique identifier is active or decommissioned, and in the latter case, the reasons for the decommissioning.

7.11. Question: What is the purpose of the exceptions laid out to in the second sentence of Article 38(1) concerning access to the information referred to in Article 33(2) and the information on the status of a unique identifier?

Answer: As explained in recital 38 of Regulation 2016/161, the purpose of those exceptions is to allow parties required to verify the authenticity of medicinal products to access the information referred to in Article 33(2) and the information on the status of a unique identifier when verifying/decommissioning the unique identifier, as such information is necessary for the proper performing of the verification/decommissioning operations.

8. OBLIGATIONS OF MARKETING AUTHORISATION HOLDERS, PARALLEL IMPORTERS AND PARALLEL DISTRIBUTORS

8.1. Question: Can marketing authorization holders delegate the performing of their obligations under Articles 40 and 41 to a third party?

Answer: Marketing authorisation holders can (but are not obliged to) delegate part of their obligations under Articles 40 and 41 to a third party by means of a written agreement between both parties. However, marketing authorisation holders remain legally responsible for those tasks.

In particular, marketing authorisation holders can delegate the performing of their legal obligation under Article 40(a) and 40(b), as well as the decommissioning task in referred to in Article 41.

8.2. Question: Situations arise where, for the same batch of product, competent authorities from different Member States issue different levels of recall, e.g. patient level Vs wholesaler level, or no recall at all. How will Article 40 of Regulation (EU) No 2016/161 work in this type of scenario?

Answer: Article 40 of Regulation (EU) No 2016/161 would not apply to recalls at patient level as the scope of the delegated act does not extend beyond the supply of the medicinal product to the end consumer. Where a medicinal product is recalled at pharmacy level in a Member State and at wholesale level in another, the marketing authorisation holder should customise the information he needs to provide in the relevant national/supranational repositories in accordance with Article 40(c).

8.3. Question: Certain Member States have national systems managing recalls and withdrawals of medicinal products in place. Would it be possible to interface those national systems with the repositories system for the verification of the safety features?

Answer: The delegated Regulation No 2016/161 does not provide for the connection between the national systems for recalls/withdrawal of medicinal product and the repositories system. Such connections may be considered by the legal entities managing the relevant repositories in the repositories system, on a voluntary basis.

9. LISTS OF DEROGATIONS AND NOTIFICATIONS TO THE COMMISSION

9.1. Question: Can marketing authorisation holders submit their proposals for amendments to Annex I of Regulation (EU) No 2016/161 to the Commission?

Answer: Only Member States notifications are taken into account for the purpose of establishing Annex I and II of the delegated Regulation, in accordance with Article 54a(2)(c) of Directive 2001/83/EC. Concerning Annex I, Member States may inform the Commission of medicinal products which they consider not to be at risk of falsification (Article 54a (4) of Directive 2001/83/EC).

9.2. Question: How should the term "Kits" referred to in Annex I to Regulation (EU) No 2016/161 be interpreted?

Answer: The term "kit" is defined in Article 1(8) of Directive 2001/83/EC. It refers to "any preparation to be reconstituted or combined with radionuclides in the final radiopharmaceutical, usually prior to its administration".

Index

2D data matrix codes xiv, 32, 148, 303

A

ABPI *see* Association of the British Pharmaceutical Industry
accuracy of data 188
ACMD *see* Advisory Council on the Misuse of Drugs
active pharmaceutical ingredients (API) 18, 241, 250, 300
active substances
 Anatomical Therapeutic Category code 226
 audits 237, 258
 authenticity checks 296
 batches 231, 238, 296
 brokering/brokers 229, 238
 calibration 232, 238
 CAPA 230
 CAS registration numbers 225
 certificate of analysis 234
 Commission publications 298
 Community Code changes xiii
 Community Database 241
 competent authorities 223, 241
 complaints 236
 compliance 244, 256, 294, 298
 conditions of registration 244
 consignee aspects 235, 238
 consignment documentation 296
 contamination 232, 238
 contracts 230
 definitions 220
 deliveries 230, 234
 deviations 230, 238
 Directive 2001/83/EC 215, 229, 240, 245, 247
 Directive 2011/62/EU 241, 246, 247
 distribution/distributors xx, 11, 215, 219, 240
 division of 229
 documentation 223, 226, 230, 245, 246, 252, 293, 300
 dosage aspects 296
 EEA States 246
 equipment 232
 EudraGMDP 241
 EudraLex Volume 4 228
 EU legislation 215
 exceptional waivers 250
 expiry dates 233, 238
 exports 223, 229, 245, 247, 294
 falsified medicinal products 233, 239, 240, 246
 finished products 255
 GDP
  EU legislation 216, 217
  importation 228, 241, 245
  preface overviews xv, xx
  UK guidance 241, 245
  UK legislation 220, 222
  wholesale distribution 103, 112
 GMP
  EU legislation 216
  importation 241, 245, 253, 295
  preface overviews xx
  UK guidance 241, 245, 253
  UK legislation 220, 222
 holding operations 229, 239
 Human Medicines Regulations 2012 219
 ICH 7 standards 295
 ICH Q9 document 229
 importation xx, 11, 215, 219, 228, 240
 information provision 224, 234
 inspections 18, 237, 242, 258, 297, 299
 inspectors 253
 instructions 230
 Internet sales 241
 introduced medicines 294
 labelling 229
 licences/licensing 221, 244
 manufacture/manufacturing xx, 11, 215, 219, 229, 240
 marketing authorisation 223
 Member States 241
 MHRA 241, 242, 258
 mutual recognition agreements 298
 National Contingency Guidance 253
 nomenclature 226
 non-compliance aspects 223, 298, 300

non-EEA countries 242, 295, 298
operations 232
orders 232
packaging 229
personnel 230, 242
premises 232
procedures 231, 239, 250, 252
procurement/procuring 229, 232, 239
product recalls 236
purification 295
Q&A document 293
Qualified Persons 257, 258
Quality Risk Management 229, 239
quality systems 229, 239
quarantine 233, 239
R&D trials 294
recalls 236
receipts 232
records 230, 231
registration 218, 221, 225, 241
re-labelling 229
re-packaging 229
retest dates 233, 239
returns 235
Schedule 7A 225
self-inspections 237
signed (signature) 239, 231
stand-alone GDP guidance 228
storage 233
suppliers 229, 239
synthesis documentation 295
templates 247, 298
third countries 223, 241, 245, 255
Title IV Directive 2001/83/EC 215
transferring information 234
transport/transportation 234, 239
UK guidance 240
UK legislation 219, 220, 222
unannounced inspections 297
validations 239, 233, 297
waivers 245, 249, 252, 300
weather conditions 232
white lists 249
WHO 295
wholesale distribution 103, 112
written documentation 223, 230, 245, 246, 252, 293
addresses/websites 19, 143, 261
administration exemptions of prescription only medicines 270
adulteration 101
Advanced Therapy Medicinal Products (ATMP) 6, 11, 60, 80, 127
advance notification 116
adverse reactions 79, 127, 180
advertisements 80, 126, 167
Advisory Council on the Misuse of Drugs (ACMD) 143
alarm settings/systems 13, 109
ALCOA applied to GDP 187
allergen extracts 61
allergic disease tests 61
ambient products 158, 161, 165
anatomical therapeutical chemicals (ATC) 60
Anatomical Therapeutic Category code 226
Anatomical Therapeutic Chemical Classification System 226
animals 11, 115, 144, 261
anti-falsification strategies 240
anti-tampering devices 111, 303
API *see* active pharmaceutical ingredients
application procedures
active substance registrations 242
brokering registration 198, 207
manufacturer's licences 70
wholesale dealer's licences 70, 114
application programming interfaces/software 313
appointing Responsible Person(s) 128
Article 46b(2) declaration in lieu of full compliance 256
Article 126a authorisation 65, 69, 121, 125
Articles 46a to 47 of Directive 2001/83/EC 216
Articles 52a to 53 of Directive 2001/83/EC 218
Articles 76 to 85b of Directive 2001/83/EC 25
Articles 80 & 85b of Directive 2001/83/EC 193
assembly aspects 220
Association of the British Pharmaceutical Industry (ABPI) 177
ATMP *see* advanced therapy medicinal products
audits 101, 237, 258
authenticity checks
active substances 296
unique identifier safety features 38, 41, 45, 58, 309
authorisations
see also Marketing Authorisation
Article 126a authorisation 65, 69, 121, 125
Blood Establishment Authorisations 11, 17
brokering/brokers 207
clinical trials 10
manufacture/manufacturing 11, 26, 84, 153
non-EEA countries 154
supply to the public 311
wholesale distribution 65, 154
authority addresses in Europe 261
automated dose dispensing 312

B

barbiturate anaesthetic agents 142
barcodes xiv, 35, 303
batches
active substances 238, 231
batch numbers 32, 79, 231, 238, 308

documentation 296
BEA *see* Blood Establishment Authorisations
best practices, temperature control 160
BGMA *see* British Generics Manufacturers Association
biological products/substances 8, 159
blood
derived products 30, 138, 144, 294
establishment 11, 17
transfusions 8
Blood Establishment Authorisations (BEA) 11, 17
bona fides establishment 172, 207, 244
breakages, GDP 101
British approved names 80, 125
British Generics Manufacturers Association (BGMA) 177
British Pharmacopoeia Commission 5
brokering/brokers
see also wholesale distribution
active substances 229, 238
application procedures 198, 207
Community Code changes xiv
competent authorities 31, 195, 197, 204
compliance 28, 210
criteria of registration 199, 208
definitions 65, 196, 204
Delegated Regulation (EU) 2016/161 xv
Directive 2001/83/EC 25, 193, 206
Directive 2011/62/EU 25, 206
documentation 28, 29, 105, 198, 205
due diligence 209
EEA States 197
emergency plans 28, 198, 204, 210
EU legislation xx, 193
exports 206, 211
falsified medicinal products 25, 204, 206
financial introductions from non-EEA countries 147
finished products 11
GDP 84, 104, 105, 197, 204, 210
holding operations 206
Human Medicines Regulations 2012 196
human use medicinal products 84, 104, 193, 196, 204
information provision 200, 209
inspections 31, 195
legislation xx, 193, 196
licensing 209
management 209
Manufacturing Authorisation 26
Member States 31, 195, 197
mutatis mutandis 195
personnel 26, 105, 205, 206
preface overviews xix, xx
procurement/procuring 206
product recalls 210
quality systems 28, 104, 204
recalls 210
records 28, 31, 195, 198, 205, 207
registration 31, 195, 197, 207
Regulation (EC) No 726/2004 31, 195, 197
risk management 28
safety features xv
storage 206
suppliers 206, 211
Title VII Directive 2001/83/EC 25, 193
UK guidance 206
UK legislation xx, 196
wholesale dealer's licences 123
Business Services Authority 140
buyer arranges main carriage 151, 152

C

calibration
active substances 232, 238
temperature monitoring 160
CAPA *see* corrective actions and/or preventative actions
carriage arrangements 151
carriers of unique identifiers 35
Case Referrals Team 13
CAS registration numbers 225
cells, safety features 60
central information repositories systems 47
centrally authorised products 178
Certificate for the importation of a pharmaceutical constituent (CPC) 12
Certificate of licensing status (CLS) 12
Certificate of manufacturing status (CMS) 12
Certificate of a pharmaceutical product (CPP) 12
certificates
see also Certificate...
of analysis 234
of Conformity CE marking 109
exports 12
GDP 11
GMP 11
of origin 149
of registration 69, 121, 125
change control 135
Chemical Abstracts Service 225
chilled, definitions 112
CHMP *see* Committee for Medicinal Products for Human Use
Clinical Practice Research Datalink (CPRD) 4
Clinical Trial Authorisations (CTA) 10
clinical trials 23
closing meetings 117
CLS *see* Certificate of licensing status
CMS *see* Certificate of manufacturing status
CMT *see* Compliance Management Team
cold chain products 156, 158

cold rooms 160
Collaborating Centre for Drug Statistics Methodology 226
commercial invoices 149
commercial refrigerators 160
Commission Delegated Regulation (EU) 2016/161 xiv, 32, 148, 305
Commission Delegated Regulation (EU) No 1252/20143 229
Commission guidelines, GDP xix
Commission on Human Medicines 5
Commission publications 298
Committee for Medicinal Products for Human Use (CHMP) 6, 17
Commodity Code 150
communications, electronic communications 66
Community Code changes to xii
Community Database xiii, 168, 241
company responses 118
competence of personnel 108
Competent Authority/Authorities
  active substances 223, 241
  brokering/brokers 31, 195, 197, 204
  continued supply 175
  Delegated Regulation (EU) 2016/161 315
  Human Medicines Regulations 2012 197
  regulatory action 173
  safety features 315
  wholesale dealer's licences 173
complaints 97, 111, 236
compliance
  active substances 244, 256, 294, 298
  brokering/brokers 28, 210
  Inspection Action Group 14
  supplier qualifications 169
  wholesale dealer's licences 120, 169
  wholesale distribution 28, 120, 169
Compliance Escalation 14
Compliance Management Team (CMT) 14
Compliance Reports 120, 244, 256
computerised systems 91
conduct, inspections 116
conformity certificates 109
consignee aspects 235, 238
consignment documentation/packing 162, 296
containers 95, 103
contamination 108, 232, 238
contemporaneous, data integrity 188
continued supply 73, 175, 176
contract acceptors 100
contract givers 100
contracting activities 85, 92, 100, 230
contrast media 61
controlled drugs 138, 147, 148, 155
control(s)
  change control 135
  environmental controls 90
  execution by lethal injection 141
  picking procedures 96
  specific medicinal products 137
  temperature 155
  wholesale distribution 90, 137
cool packs 104
Co-Rapporteur 5
corrective actions and/or preventative actions (CAPA) 85, 136, 230
Council of Europe 241
counterfeits *see* falsified medicinal products
Courts, MHRA's Enforcement Group 13
CPC *see* Certificate for the importation of a pharmaceutical constituent
CPP *see* Certificate of a pharmaceutical product
CPRD *see* Clinical Practice Research Datalink
CPS *see* Crown Prosecution Service
criminal activities 13, 55, 101
criteria of brokers registration 199, 208
critical deficiencies, inspections 117
critical findings, IAG 15
criticality of data 186
cross-border verification 314
cross-contamination *see* contamination
Crown Prosecution Service (CPS) 13
CTA *see* Clinical Trial Authorisations
customers
  due diligence 171
  qualification(s) 95, 168, 170, 171
customs procedures 147, 149
cytostatic materials 112
cytotoxic materials 137

D

data
  ALCOA applied to GDP 187
  databases xiii, 168, 241
  matrix codes xiv, 32, 148, 303
  ownership 54, 314
  protection 54
  routers (hubs) 47, 49, 313
Data Governance 186, 189
data integrity (DI) 185
Data Processing Group 10, 11
dealer's licences *see* wholesale dealer's licences
declaration in lieu of full compliance 256
Declaration Unique Consignment Reference (DUCR) 150
decommissioning aspects 38, 42, 309
Defective Medicines Report Centre (DMRC) 12, 18, 179, 180
defective products 179, 180
deficiencies, inspections 117
degradation considerations 156
Delegated Regulation (EU) 2016/161 xiv, 32, 148, 303

anti-tampering devices 111, 303
applicable to 33
brokering/brokers xv
central information repositories systems 47
competent authorities 315
data routers (hubs) 47, 49, 313
decommissioning 38, 42, 309
definitions 33, 34
derogations lists 41, 44, 57, 316
Directive 2001/83/EC xiv, 32, 148, 303
Directive 2011/62/EU 32
distributors 55, 315
entry into force 59
equivalent unique identifiers 40, 313
falsified medicinal products 32, 148, 310
free samples 55
hub repositories system 47, 49, 313
human use medicinal products xiv, 32, 148, 305
importation 55, 315
individual pack identification 58
marketing authorisations 55, 305, 315
National Competent Authorities 33, 54, 56, 310
national repositories 47, 313
notifications 57, 316
packaging xiv, 32, 148, 303
parallel distributors 55, 315
parallel importers 55, 315
personnel xv, 43, 311
product recalls 55, 315
Qualified Person(s) xv
recalls 55, 315
repositories system 46, 312
Safety Feature Legislation xiv, 32, 148, 303
stolen products 55
subject matter 33
supranational repositories 47, 313
tampering xv, 32, 41, 43, 303
technical specification 34, 306
transitional safety measures 58
unique identifiers xiv, 32, 148, 303
verification 38, 309
wholesale distribution xv, 41
withdrawn products 55, 315
Delegated Regulation (EU) No 1252/20143 229
delivery considerations 230, 234
Department of Health and the Care Quality Commission 140
Department of Health (DH) 177
derogations lists, Delegated Regulation (EU) 2016/161 41, 44, 57, 316
destruction procedures 96, 137
deviation management 109, 136, 230, 238
devices, MHRA innovation 6, 7
DH *see* Department of Health
DI *see* data integrity
diluting agents 60
Directive 2001/83/EC
active substances 215, 229, 240, 245, 247
as amended, Title VII xix, 25, 128, 145, 193
Articles 46a to 47 216
Articles 52a to 53 218
brokering/brokers 25, 193, 206
continued supply 177
Delegated Regulation (EU) 2016/161 xiv, 32, 148, 305
exports 145
GDP xix, 30, 83
glossary of legislation xxii
Human Medicines Regulations 2012 65
human use medicinal products 83, 193
importation 55, 215, 315
manufacture 215
Responsible Person(s) 128
Title III 164, 258
Title IV 215
Title VII xix, 25, 128, 145, 193
wholesale dealer's licences xv, 25, 128, 145
wholesale distribution 25, 65, 128, 145, 177, 193
Directive 2011/62/EU (Falsified Medicines)
active substances 241, 246, 247
brokering/brokers 25, 206
Community Code changes xii, xiv
Delegated Regulation (EU) 2016/161 32
GDP 9, 83
preface overviews xix
registrations 11
Safety Features Legislation 32
wholesale distribution 25, 145
Directives
see also Directive 2001/83/EC; Directive 2011/62/EU
2002/98/EC 65
2003/94/EC xxii
2004/23/EC 65
2004/27/EC 177
EDQM 8
discontinued products 177
dispensing opticians 269
disposal procedures 96, 137, 144
distribution/distributors
see also Good Distribution Practice; wholesale distribution
active substances xx, 11, 215, 219, 240, 241
definitions 238
Delegated Regulation (EU) 2016/161 55, 315
EU legislation 215
Human Medicines Regulations 2012 219
legislation 215, 219
safety features 55, 315
specific medicinal products 144
UK guidance 240, 241
UK legislation 219
unique identifiers 55, 315
wholesale dealer's licences 121

diverted medicines 173, 229
DMRC *see* Defective Medicines Report Centre
documentation
  see also written documentation
  active substances 223, 226, 230, 245, 246, 252, 293, 300
  APIs 300
  brokering/brokers 28, 29, 105, 198, 205
  GDP 92, 105, 110, 202, 205, 230
  Human Medicines Regulations 2012 73, 76, 223, 226
  human use safety features 293
  ICH Q9 document 86, 163, 229
  inspections 297, 299
  instructions 92, 230
  intermediate products 295
  language 203
  personnel 203
  procedures 231, 239, 250, 252
  Q&A's 108, 293, 302
  quality systems 202
  retention of 203
  starting materials 295
  synthesis 295
  templates 247, 255, 298
  unique identifiers 39
  waivers 245, 249, 252
  wholesale dealer's licences 73, 76, 122, 124
  wholesale distribution 28, 29, 73, 76, 92, 122, 124, 149
dosage aspects 296
Drug Alert issuance 181
drug/device combinations 6
drug misuse 139, 143, 155
Drug Statistics Methodology 226
drug substances (active pharmaceutical ingredients) 18, 241, 250, 300
DUCR *see* Declaration Unique Consignment Reference
due diligence 127, 169, 171, 209
duration of licences 72
duties of Responsible Person(s) 128
Duty Compliance Officers 143

E

EC *see* European Community
ECO *see* Export Control Organisation
Economic Operator Registration and Identification (EORI) 150
EDQM *see* European Directorate for Quality of Medicines
EEA *see* European Economic Area
electrolyte balance 60
electronic communication 66
EMA *see* European Medicines Agency
embargoes 141
emergency plans
  brokering/brokers 28, 198, 204, 210
  wholesale dealer's licences 73, 122
  wholesale distribution 28, 73, 122
enforcement authorities 7, 13, 15
entitled persons to supply to the public 311
entry into force 59
environmental controls 90
EORI *see* Economic Operator Registration and Identification
ephedrine 138
equipment
  active substances 232
  alarm settings/systems 13, 109
  calibration 160
  CE certificate of Conformity 109
  conditions of holding licences 121
  Directive 2001/83/EC 27
  GDP 88, 90, 108, 232
  temperature considerations 109, 160
  wholesale dealer's licences 73, 121
equivalent licenced product availability 179
equivalent standards 297
equivalent unique identifiers 40, 313
establishment
  blood establishment 11, 17
  bona fides 172, 207, 244
  repositories system 46, 312
EU *see* European Union
EudraGMDP xiii, 168, 241
EudraLex Volume 4 228
EU Referendum xi
European authority addresses 261
European Commission 108
European Community (EC) *see* Regulation (EC)
European Directorate for Quality of Medicines (EDQM) 8
European Economic Area (EEA)
  active substances 246
  brokering/brokers 197
  GDP 9
  Human Medicines Regulations 2012 66, 74
  parallel distribution 175
  parallel importers 174
  transactions to non-EEA countries 153
  unlicenced medicinal products 12
  wholesale dealer's licences 74, 122, 124
  wholesale distribution 66, 74
European Medicines Agency (EMA)
  exports to non-EEA countries 148
  GMP 8
  Inspection Action Group 17
  parallel distribution 175
  parallel importers 174
  scientific advice 6
  unlicenced medicinal products 178
  wholesale dealer's licences 74, 125

INDEX

ALCOA applied to 187
anti-tampering devices 111
audits 101, 237
breakages 101
brokering/brokers 84, 104, 105, 197, 204, 210
CAPA 85
CE certificates of Conformity 109
certificates 11
Commission guidelines xix
Community Code changes xiii
complaints 97, 111, 236
compliance 169
computerised systems 91
containers 95, 103
contamination 108
contract acceptors 100
contract givers 100
contracting activities 85, 92, 100
contracts 92
cool packs 104
data integrity 185
definitions 107
deliveries 230, 234
destruction procedures 96
deviation aspects 109
Directive 2001/83/EC xix, 30, 83
Directive 2011/62/EU 9, 83
documentation 92, 105, 110, 202, 205, 230
environmental controls 90
equipment 88, 90, 108, 232
EudraLex Volume 4 228
EU guidelines 9, 82, 202
EU legislation 216, 217
exports 83, 97
falsified medicinal products 9, 97, 99, 107, 111, 172, 233, 239
free zones/free warehouses 84
guideline preface overviews xix
holding conditions/operations 83, 121
Human Medicines Regulations 2012 9, 76
human use medicinal products 82, 202, 228
importation 89, 93, 228, 241, 245
information provision 234
Inspection Action Group 17
inspections 17, 101, 237
instructions 92
insulated boxes 104
investigation procedures 110
labelling 103
legislation 216, 217, 220, 222
Manufacturing Authorisation 84
marketing authorisation holders 93
MHRA 8, 9
narcotics 103
non-EEA States 9
operations 93, 110, 232
orders 232
outsourced activities 85, 100
packaging 103
personnel 86, 105, 230
picking procedures 96
preface overviews xix
premises 88, 108, 232
procurement/procuring 83, 94, 229, 232, 239
product recalls 97, 99, 111, 236
products requiring special conditions 103
psychotropic substances 103
qualification(s) 91
Quality Management 84
Quality Risk Management 86
quality systems 84, 104, 229, 239
radioactive materials 89, 103
recalls 97, 99, 111, 236
receipts 95, 232
records 92, 203, 205, 230, 231
Responsible Person(s) 86
returns 97, 98, 235
risk-based inspections 120
safety features 110
sales representative sample goods 167
security 91, 110
self-inspections 101, 237
specific medicinal products 144
stand-alone guidance 228
stolen goods 101
storage 83, 89, 95, 102, 233
suppliers/supplies 83, 94, 96
tampering 111
temperature considerations 90, 103, 103, 109, 112
temperature controls 90, 103, 109, 112
theft 101
third countries 89, 97
third party transportations 102
Title VII Directive 2001/83/EC xix
transportation 101, 112, 234, 239
transport/transportation 101, 112, 234, 239
UK guidance 115
UK legislation 220, 222
unusual sale patterns 110
validation 91
weather conditions 89, 109
wholesale dealer's licences 76, 115, 172
wholesale distribution 76, 83, 92, 115, 137, 172
written documentation 92, 203, 293
Good Laboratory Practice (GLP) 8, 10
Good Manufacturing Practice (GMP)
active substances
EU legislation 216
exports 295
importation 241, 245, 253, 295
legislation 216, 220, 222
preface overviews xx
UK guidance 241, 245, 253
UK legislation 220, 222

certificates 11
equivalent standards 297
exports 295
guidelines xiii
importation 241, 245, 253, 295
Inspection Action Group 17
legislation 216, 220, 222
MHRA Inspectorate Group 8
non-EEA countries 153
standards 297
UK legislation 220, 222
wholesale distribution 153
Good Pharmacovigilance Practice (GPvP) 8, 10, 17
Good Practice Standards (GxP) 16, 119, 185
GPvP *see* Good Pharmacovigilance Practice
granting licences 68, 115
graphical user interfaces (GUI) 313
GSL *see* General Sale Lists
GUI *see* graphical user interfaces
GxP *see* Good Practice Standards

H

handling procedures 121, 164
hard copy records 186
hazardous products 89, 138
health care professional's supply restrictions 139
health certificates 149
health protection 311
herbal products/substances 66, 68, 121, 125
highly active materials 112, 137
HMRC exports 150
holding operations
active substances 229, 239
brokering/brokers 206
definitions 107
GDP 83, 121
wholesale dealer's licences 72, 121
wholesale distribution 154
Home Office 143
Home Office Drug Licensing and Compliance Unit 138
homeopathic medicinal products 30, 60, 67
hospital supplies 311
hub repositories system 47, 49, 313
Human Medicines Regulations 2012
active substances 219
Article 126a authorisation 65
brokering/brokers 196
citations 65, 196, 219
commencement 65, 196, 219
communication 66
competent authorities 197
Data Processing Group/MHRA 12
Directive 2001/83/EC 65
Directive 2002/98/EC 65
Directive 2004/23/EC 65
distribution 219
documentation 73, 76, 223, 226
duration of licences 72
EEA States 66, 74
electronic communication 66
exempt ATMPs 80
exports 66, 75, 223
falsified medicinal products 66, 75
Fees Regulations 66
GDP 9, 76
Gold Standard 130
granting licences 68
herbal products/substances 66
homeopathic medicinal products 67
importation 67, 75, 219
information provision 200, 224
inspectors 67
interpretations 65, 196, 220
licences/licensing 67, 68, 221
manufacture 219
marketing authorisation 66, 73, 75
MHRA 5
personnel 267
persons who can be supplied to 267
preface overviews xix, xxii
recalls/withdrawals 180
records 74
refusal of licences 71
registrations 69, 197, 221, 225
Regulation (EC) No 726/2004 67
Regulation (EC) No 1234/2008 67
Regulation (EC) No 1394/2007 67
Responsible Persons(s) 77, 130
Schedule 4 standard provisions of licences 72, 78
Schedule 6 Manufacturer's and Wholesale Dealer's Licences 80
Schedule 17 provisions 269
standard provisions of licences 72, 78
UK legislation 64
unlicenced medicinal products 178, 179
wholesale dealer's licences 130, 178
wholesale distribution 64, 130, 178, 267
human readable code/information xv, 32, 37, 307
human use medicinal products
brokering/brokers 193, 196
Delegated Regulation (EU) 2016/161 xiv, 32, 148, 305
Directive 2001/83/EC xiv, 32, 148, 193, 305
EU legislation 215
European authority addresses 261
GDP 82, 202, 228
Q&A documents 293, 302
safety features 32, 110, 305
UK legislation 219
wholesale dealer's licences 115
hygiene, personnel 88, 108

I

IAG *see* Inspection Action Group
ICH 7 standards 295
ICH Q9 document 86, 163, 229
immunological medicinal products 30, 138
IMP *see* Investigational Medicinal Products
importation/importers/imports
  see also exports
  active substances xx, 11, 215, 219, 228, 240, 293
  APIs 241
  blood-derived products 294
  definitions 220
  Delegated Regulation (EU) 2016/161 55, 315
  Directive 2001/83/EC 55, 215, 315
  EU legislation 215
  flowcharts for APIs 250
  GDP 89, 93, 228, 241, 245
  GMP 241, 245, 253, 295
  Human Medicines Regulations 2012 67, 75, 219
  introduced medicines 289, 294
  legislation 215, 219
  non-EEA countries 145, 153
  plasma-derived products 294
  Q&A document 293
  safety features 55, 315
  third countries 30
  Title IV Directive 2001/83/EC 215
  UK guidance 145, 153, 173, 240
  UK legislation 219
  UK licensing 289
  unique identifiers 55, 315
  unlicenced medicinal products 12, 178
  wholesale dealer's licences 75, 122, 126, 146
  wholesale distribution 67, 75, 145, 153, 173, 178, 289
Incoterms® 2010 rules 151
individual pack identification 58
information provision
  active substances 224, 234
  brokering/brokers 200, 209
  GDP 234
  Human Medicines Regulations 2012 200, 224
  repositories system 47, 56
  sources of 287
  wholesale dealer's licences 125
initial temperature mapping 156
injections 141, 144
Innovation Office, MHRA 6
INS Data Processing Group 12
Inspection Action Group (IAG) 11, 14, 15, 16
Inspection, Enforcement and Standards Division 7, 15
inspections
  active substances 18, 237, 242, 258, 297, 299
  brokering/brokers 31, 195
  closing meetings 117
  company responses 118
  Compliance Reports 120
  conduct 116
  critical deficiencies 117
  deficiencies 117
  documentation 297, 299
  fees 115
  GDP 17, 101, 237
  GxP 119
  introductory/opening meetings 116
  major deficiencies 117
  meetings 116
  MHRA 242, 258
  non-critical deficiencies 117
  notification 116
  opening meetings 116
  planning 115
  reports 118
  risk-based 118
  risk rating 120
  self-inspections 101, 237
  Sentinel risk information module 119
  summary/closing meetings 117
  wholesale dealer's licences 115
Inspectorate Group 8
inspectors 67, 253
instructions 92, 230
insulated boxes 104
integrity, data integrity 185
Intelligence Analysts 13
intermediate products 295
international non-proprietary names 80, 125
International Union of Pure and Applied Chemistry nomenclature 226
Internet
  active substances 241
  Data Processing Group registrations 11
  websites 19, 143, 261
intravenous solution additives 60
introduced medicinal products 146, 289, 294
introductory/opening meetings 116
Investigational Medicinal Products (IMP) 11, 18, 68
investigation procedures 110
Investigations Team 13
irrigating solutions 60

K

ketamine reclassification 142
kit safety features 60, 316

L

labelling 103, 229
laboratories, good practice 8, 10
language usage 203
large refrigerators 160

legibility of data 187
legislation
  see also Directives
  active substances 215, 219, 245
  brokering/brokers xx, 193, 196
  clinical trials 23
  distribution 215, 219
  European Union xx, xxii, 23, 193, 215, 216, 217
  fees 23
  GDP 216, 217, 220, 222
  glossary of xxii
  GMP 220, 222
  human use product safety features xiv, 32, 46, 148, 305
  importation 215, 219
  Inspectorate Group 18
  manufacture 215, 219
  persons who can be supplied to 268
  preface overviews xii
  registration 218
  repositories system 53
  safety features xiv, 32, 46, 148, 303
  UK guidelines 64
  wholesale distribution 23, 64
lethal injection drug controls 141
licences/licensing
  see also manufacturer's licences; wholesale dealer's licences
  active substances 221, 244
  application procedures 114
  brokering/brokers 198, 209
  controlled drugs 138
  duration of 72
  execution by lethal injection 141
  Exempt ATMPs 11
  exports into the UK 289
  GMP Inspectors 8
  holding conditions 123
  Human Medicines Regulations 2012 67, 68, 221
  importation into the UK 289
  IMPs 11
  Inspectorate Group 18
  introduced medicines 289
  precursor chemicals 138
  refusals 71, 115
  revoking licences 74, 123, 127
  "Specials" Licences 8, 11, 12, 148, 153
  statement of status 12
  suspensions 74, 123, 127
  veterinary products 11, 115
  wholesale distribution 67, 68, 114, 289
lisdexamfetamine controls 142

M

main carriage arrangements 151
major deficiency considerations 117
management
  see also Quality Risk Management
  brokering/brokers 209
  deviation management 109, 136, 230, 238
  Quality Management 84, 132
  recalls 181
  repositories system 53, 312
  reviews/monitoring 85, 136
mandatory requisition forms 139
manufacture/manufacturing
  see also Good Manufacturing Practice; manufacturer's
  active substances xx, 11, 215, 219, 229, 240
  APIs 241
  Authorisation 11, 26, 84, 153
  decommissioning unique identifiers 310
  definitions 221
  Directive 2001/83/EC 215
  EU legislation 215
  Human Medicines Regulations 2012 219
  legislation 215, 219
  Title IV Directive 2001/83/EC 215
  UK guidance 219, 240, 241
  UK legislation 219
  unique identifiers 39, 310
manufacturers, unique identifiers 39, 310
Manufacturer's/Importer's Authorisation (MIA)
  active substances 254
  Data Processing Group/MHRA 11
  non-EEA countries 153
  parallel distribution 175
  parallel importers 174
  wholesale distribution 153
manufacturer's licences
  application procedures 70
  Data Processing Group/MHRA 11
  holding conditions 123
  Schedule 6 Licences 80
  "Specials" Licences 8, 11, 12, 148, 153
manufacturer's unique identifiers 39, 310
mapping temperature 90, 103, 109, 156
Marketing Authorisation
  active substances 223
  Community Code changes xiv
  Delegated Regulation (EU) 2016/161 55, 305, 315
  GDP 93
  holders xiv, 55, 93, 315
  Human Medicines Regulations 2012 66, 73, 75
  safety features 55, 305, 315
  unique identifiers 55, 305, 315
  unlicenced medicinal products 12, 178
  wholesale dealer's licences 73, 75, 121, 124, 178
  wholesale distribution 66, 73, 75
matrix codes xiv, 32, 148, 303
Mean Kinetic Temperature (MKT) 163
medicinal products, definitions 65

Medicines and Healthcare products Regulatory Agency (MHRA) 3
  active substances 241, 242, 258
  addresses/contact details 19
  Community Code changes xiii
  continued supply 175
  data integrity 185
  Data Processing Group 10
  devices sector 7
  DMRC 12, 18
  enforcement 7, 13, 15
  GCP 8, 10
  GDP 8, 9
  GLP 8, 10
  GMP 8
  GPvP 8, 10
  Innovation Office 6
  Inspection Action Group 11
  Inspection, Enforcement and Standards Division 7, 15
  inspections 115, 242, 258
  Inspectorate Group 8
  overview 5
  Pharmacovigilance Inspectorate 10
  Process Licensing Portal 114
  registrations 11, 241
  risk-based inspection 118
  standards 7, 15
  supplier qualifications 168
  unlicenced medicinal products 178
  websites 19
  wholesale dealer's licences 114, 125
Medicines for Human Use (Clinical Trials) Regulations 2004 23
Medicines (Products for Human Use) (Fees) Regulations 2016 23
meetings, inspections 116
Member States
  active substances 241
  brokering/brokers 31, 195, 197
  continued supply 176
  Delegated Regulation (EU) 2016/161 42, 304
  Directive 2001/83/EC 25, 42, 304
  safety features 42, 304
  supply chains 42
MHRA *see* Medicines and Healthcare products Regulatory Agency
MIA *see* Manufacturer's/Importer's Authorisation
Misuse of Drugs Act 1971 143, 155
Misuse of Drugs Regulations 2001 139
MKT *see* Mean Kinetic Temperature
monitoring
  management 85, 136
  temperature 155, 159, 160
monograph names 80, 125
MRA *see* Mutual Recognition Agreements
mutatis mutandis 195
Mutual Acceptance of Data 10
Mutual Recognition Agreements (MRA) 298

N

name consideration 80, 125, 226
nanotechnology 6
narcotics 30, 103, 138, 233
National Competent Authorities 33, 54, 56, 310
National Contingency Guidance 253, 255
National Institute for Biological Standards and Control (NIBSC) 4
National Institute for Health and Care Excellence (NICE) 6
national reimbursement numbers 307
national repositories 47, 313
NHS Business Services Authority (NHSBSA) 140
NIBSC *see* National Institute for Biological Standards and Control
NICE *see* National Institute for Health and Care Excellence
non-compliance considerations 223, 298, 300
non-critical deficiencies 117
non-defective products 164
non-EEA countries
  active substances 242, 295, 298
  authorisations 154
  exports 25, 145
  falsified medicinal products 145
  GDP 9
  GMP 153
  importation 145, 153
  introduced medicinal products 146
  Manufacturer's/Importer's Authorisation 153
  manufacturer's specials licences 148, 153
  procurement/procuring 145, 154
  qualification checks 150
  restrictions 149
  transactions to EEA States 153
  unlicenced medicinal products 12
  wholesale dealer's licences 25, 124, 146
  wholesale distribution 124, 145, 146, 150
non-proprietary names 80, 125
notification considerations 57, 116, 316
novel drug/device combinations 6
nutrition solutions 60

O

obsolete goods (destruction of) 96
OECD *see* Organisation for Economic Co-operation and Development
offence investigations 13, 55, 101
opening meetings 116
operations
  active substances 232

best practices in unusual sale patterns 110
GDP 93, 110, 232
repositories system 52
separately: definition of 110
unusual sale patterns 110
optometrist's supply restrictions 269
orders 232
Organisation for Economic Co-operation and Development (OECD) 10
original records (true copies) 188
osmotic diuresis 60
outsourced activities 85, 100
ownership, data ownership 54, 314

P

packaging
active substances 229
barcodes xiv, 37, 303
Delegated Regulation (EU) 2016/161 xiv, 32, 148, 303
GDP 103
safety feature xiv, 32, 148, 303
parallel distribution 55, 175, 315
parallel importers 55, 173, 315
Parallel Import Licensing Scheme 173
parenteral nutrition solutions 60
Parliamentary Orders 142
partial package supplies 45
periodic reviews 108
permanent data, ALCOA applied to GDP 187
Person Appointed Hearings 19
personnel
see also Qualified Person(s); Responsible Person(s)
active substances 230, 242
ALCOA applied to GDP 187
authorised/entitled to supply to the public 311
brokering/brokers 26, 105, 205, 206
competence 108
conditions of holding licences 121
decommissioning unique identifiers 311
documentation 203
GDP 86, 105, 230
Human Medicines Regulations 2012 267
hygiene 88, 108
periodic reviews 108
responsibilities 8, 128
specific medicinal products 138
those who can be supplied to 267
training 88, 108
unique identifiers xv, 43, 311
wholesale dealer's licences 73, 121
wholesale distribution 26, 73, 86, 121, 267
pharmaceutical refrigerators 158
Pharmacovigilance Inspectorate 10
pharmacy medicine guidelines 268
physical introductions from non-EEA countries 147
picking procedures 96
planning inspections 115
plasma-derived products 294
PLPI *see* Product Licence for Parallel Import
Policy/Relationships management 13
practical experience considerations 129
precursor chemical controls 138
pre-distribution partners 153
premises
active substances 232
Directive 2001/83/EC 27
GDP 88, 108, 232
wholesale distribution 73, 121, 125
prescription only medicines
administration exemptions 270
exemptions 270
persons who can be supplied to 268
sale or supply 270
preventative actions 85, 136, 230
price harmonisation 174
printing quality 36, 307
probes for temperature monitoring 109, 160
procedure documentation 231, 239, 250, 252
Process Licensing Portal 114
procurement/procuring
active substances 229, 232, 239
brokering/brokers 206
definitions 107
GDP 83, 94, 229, 232, 239
non-EEA countries 145, 154
specific medicinal products 144
third countries 170
wholesale distribution 144, 145, 154
product codes 32, 308
product discontinuations 177
Product Licence for Parallel Import (PLPI) 174
Product Licences 12
product proof of release 95, 110
product recalls
active substances 236
brokering/brokers 210
Delegated Regulation (EU) 2016/161 55, 315
GDP 97, 99, 111, 236
parallel importers 174
safety features 55, 315
unique identifiers 55, 315
wholesale distribution 174, 180
products requiring special conditions 103
product withdrawals 55, 180, 315
prohibition considerations 149
proof of release 95, 110
Prosecution Units 13
protection
data protection 54
public health protection 311

provision of information *see* information provision
pseudoephedrine 138
psychotropic substances 30, 103, 138
public health protection 311
purification 295

Q

Q&A documents 108, 293, 302
QP *see* Qualified Persons
QRM *see* Quality Risk Management
qualification(s)
  customers 95, 168, 170, 171
  definitions 107
  GDP 91
  non-EEA countries 150
  suppliers/supplies 94, 168
  trading partners 95, 168, 170
  wholesale distribution 168, 170
Qualified Person(s) (QP)
  active substances 17, 257, 258
  exports 148
  Inspection Action Group 17
  preface overviews xiii
  safety features xv
  wholesale distribution 148
quality defects 180
Quality Management 84, 134
Quality Risk Management (QRM)
  active substances 229, 239
  data integrity 186
  definitions 107
  GDP 86
  temperature monitoring 163
  wholesale distribution 135, 163
quality systems
  active substances 229, 239
  brokering/brokers 28, 104, 204
  contracting activities 85
  definitions 107
  documentation 202
  GDP 84, 104, 229, 239
  management reviews/monitoring 85, 136
  outsourced activities 85
  wholesale dealer's licences 75, 123
  wholesale distribution 28, 84, 135
quarantine procedures/status 149, 233, 239
question & answer (Q&A) documents 108, 293, 302

R

R&D trials 294, 304
radioactive materials 89, 103
radionuclide generators 60
radionuclide precursors 60
radiopharmaceuticals 30, 68, 137, 144, 316
Rapid Alert System 13
Rapporteur 5
rating process, risk 120
raw materials 127
recalls
  active substances 236
  brokering/brokers 210
  Delegated Regulation (EU) 2016/161 55, 315
  follow-up action 184
  GDP 97, 99, 111, 236
  issuing procedures 181
  parallel importers 174
  safety features 55, 315
  test processes 183
  unique identifiers 55, 315
  wholesale distribution 174, 180
receipts 95, 158, 232
records
  active substances 230, 231
  brokering/brokers 28, 31, 195, 198, 205, 207
  data integrity 186
  GDP 92, 203, 205, 230, 231
  Human Medicines Regulations 2012 74
  special medicinal products 125
  unique identifiers 39
  wholesale dealer's licences 74, 122, 124
  wholesale distribution 28
Reference Member State (RMS) 176
Referendum (EU) xi
referral considerations 15
refreshment rooms 90
refrigerators 158
refusal of licences 71, 115
refuse procedures 96, 137, 144
registration
  active substances 218, 221, 225, 241
  application procedures 198
  brokering/brokers 31, 195, 197, 207
  Community Database 241
  conditions of 244
  criteria 199, 208
  Data Processing Group, MHRA 11
  Directive 2011/62/EU 11
  EU legislation 218
  Human Medicines Regulations 2012 69, 197, 221, 225
  importation/manufacture 218
  legislation 218
  MHRA 11, 241
  requirements for 222, 225
  wholesale distribution 69
Regulation (EC) No 726/2004 25, 31, 67, 195, 197
Regulation (EC) No 1234/2008 67
Regulation (EC) No 1394/2007 67
Regulations 220 & 221 exemptions 279

regulatory considerations 6, 173
  see also Delegated Regulation..; Directives; Regulation...
re-labelling 229
Relationships management 13
release, proof of 95, 110
relevant EU provisions, definitions 67
relevant European State, definitions 67
relevant medicinal products, definitions 67
removing safety features/unique identifiers 40, 55
re-packaging 229
replacing safety features 40
reports 118
repositories system 46, 312
  accessibility 312
  characteristics of 50
  data ownership/protection 54
  Delegated Regulation (EU) 2016/161 46, 312
  establishment 46, 312
  functioning of 49
  information provision 47, 56
  legislation 53
  management 53, 312
  National Competent Authorities 54, 56
  operations of 52
  removing unique identifiers 55
  safety features 46, 312
  structure of 47
  supervision 56
  uploading information 47
re-qualification considerations 169, 171
requisition forms 139
research and development trials (R&D) 294, 304
responsibilities
  GMP Inspectors 8
  Responsible Person(s) 128
Responsible Person(s) (RP)
  appointment of 128
  Directive 2001/83/EC 128
  duties 128
  experience 129
  GDP 86
  Gold Standard 130
  Human Medicines Regulations 2012 77, 130
  Inspection Action Group 17
  preface overviews xix
  Quality Management 136
  responsibilities 128
  specific medicinal products 137
  Title VII Directive 2001/83/EC 128
  wholesale dealer's licences 77, 121, 125, 130
  wholesale distribution 136
rest rooms 90
retail sale or supply 68, 124, 268
retaining documentation 203
retest dates 233, 239
returns
  active substances 235
  ambient products 165
  GDP 97, 98, 235
  non-defective products 164
  refrigerated products 164, 166
  unlicenced sites 165
  wholesale dealer's licences 165
  wholesale distribution 164
reviews 85, 108, 136
revoking licences 74, 123, 127
risk
  see also Quality Risk Management
  Estimation Tool 119
  information module 119
  inspection programmes 120
  management 28
  rating process 120
  temperature monitoring 163
routine re-qualification 169, 171
RP *see* Responsible Person(s)

S

safety
  anti-tampering devices 111, 303
  brokering/brokers xv
  competent authorities 315
  decommissioning 38, 42, 309
  Delegated Regulation (EU) 2016/161 xiv, 32, 148, 303
  Directive 2001/83/EC xiv, 32, 148, 305
  falsified medicinal products 32, 148, 310
  features legislation xiv, 32, 148, 303
  GDP 110
  human use medicinal products 32, 110, 305
  kits 316
  legislation xiv, 32, 148, 303
  packaging xiv, 32, 148, 303
  personnel xv, 43, 311
  prescription only medicines 268, 270
  Q&A documents 302
  radiopharmaceuticals 316
  removal/replacement 40, 55
  special medicinal products 124
  tampering xiv, 32, 41, 43, 303
  transitional measures 58
  unique identifiers xiv, 32, 148, 303
  verification 38, 309
  wholesale dealer's licences xv, 75
  wholesale distribution xv, 41, 75
sales
  exemptions 270
  prescription only medicines 268, 270
  proof of release 95, 110
sales representative samples 167

sanctions 141
SAWP *see* Scientific Advice Working Party
Schedule 2 & 3 controlled drugs 139
Schedule 4 standard provisions of licences 72, 78
Schedule 6 Manufacturer's and Wholesale Dealer's Licences 80
Schedule 7A – information provision 225
Schedule 17 provisions 269
scientific advice
  from the EU 6
  from the UK 6, 7
Scientific Advice Working Party (SAWP) 6
security
  computerised systems 91
  GDP 91, 110
segregation considerations 108
self-inspections 101, 237
seller arranges main carriage 152
Sentinel risk information module 119
separately, definitions 110
serial numbers 32, 308, 313
Shipman Inquiry 139
short term storage 166
signed (signature) 239, 231
site inspections 116
small refrigerators 159
software packages 313
sole UK pre-distribution partners 153
solutions, safety features 60
solvents, safety features 60
special medicinal products 78, 125
  see also unlicenced medicinal products
  advertisements 126
  documentation 124, 125
  importation 126
  MHRA 125
  records 125
  safety 124
"Specials" Licences 8, 11, 12, 148, 153
specific medicinal products
  controls 137, 137
  Regulations 220 & 221 279
  supply restrictions 137
specified person transactions 75
stand-alone GDP guidance 228
standard provisions of licences 72, 78, 125, 127
standards
  equivalent standards 297
  GMP 297
  Gold Standard 130
  Good Practice Standards 16, 119, 185
  ICH 7 standards 295
  Inspection, Enforcement and Standards Division 7, 15
  MHRA 7, 15
  NIBSC 4
starting materials 127, 295
statement of status 12
status reversals of unique identifiers 39
stolen goods 55, 101
storage
  active substances 233
  ambient products 158
  brokering/brokers 206
  GDP 83, 89, 95, 102, 233
  refrigerated products 158
  specific medicinal products 144
  temperature requirements 156, 160
  wholesale dealer's licences 73, 121, 166
  wholesale distribution 156, 160
strategic goods considerations 141
stratified medicines, overview 6
structure of repositories systems 47
submission templates 255
summary meetings 117
supervision, repositories system 56
suppliers/supplies
  active substances 229, 239
  brokering/brokers 206, 211
  definitions 107
  due diligence 169
  GDP 83, 94, 96
  Human Medicines Regulations 2012 68
  qualifications 168
  re-qualification 169
  restrictions 137, 180
  routine re-qualification 169
  third countries 170
  wholesale distribution 137, 155, 168, 180
supply chains 42
supranational repositories 47, 313
suspected defects 185
suspension of licences 74, 123, 127
synthesis documentation 295

T

tampering
  GDP 111
  unique identifiers xv, 32, 41, 43, 303
technical specifications of unique identifiers 34, 306
temperature
  best practices 160
  calibration 160
  consignment packing 162
  control strategies 90, 103, 109, 112, 155
  mapping 90, 103, 109, 156
  monitoring 155, 159, 160
  probes 109, 160
  sensitive products 103, 138
  storage requirements 156, 160
  transportation requirements 159, 162
  wholesale distribution 90, 103, 109, 112, 138, 155

test processes 61, 183, 233, 239
theft 55, 101
thiopental controls 142
third countries
  active substances 223, 241, 245, 255
  definitions 68
  GDP 89, 97
  imports 30
  procurement/procuring 170
  supplies to 170
  wholesale dealer's licences 75
third party considerations 102, 164, 258
thymoxamine hydrochloride 269
tissue, safety features 60
Title III, Directive 2001/83/EC 146
Title IV, Directive 2001/83/EC 215, 215
Title VII, Directive 2001/83/EC xix, 25, 128, 145, 193
traceability 127
trademark aspects 13, 125
trading partners 95, 168, 170
traditional herbal products/substances 66, 68, 121, 125
training personnel 88, 108
tramadol controls 142
transfusions, blood 8
transitional safety measures 58
Transmissible Spongiform Encephalopathies (TSE) 23
transport/transportation
  active substances 234, 239
  consignment packing 162
  definitions 107
  GDP 101, 112, 234, 239
  Incoterms® 2010 rules 151
  temperature requirements 159, 162
  wholesale distribution 101, 112, 151, 159
true copies (original records) 188
TSE *see* Transmissible Spongiform Encephalopathies
two dimensional (2D) data matrix codes xiv, 32, 148, 303

U

UK *see* United Kingdom
UN *see* United Nations
unannounced inspections 297
unauthorised medicinal products 73, 122
unique identifiers xiv, 32, 148, 303
  additional information 37
  authenticity checks 38, 41, 45, 58, 309
  batch numbers 32, 308
  carriers 35
  central information repositories systems 47
  composition 34
  data routers (hubs) 47, 49, 313
  decommissioning 38, 42, 309
  Delegated Regulation (EU) 2016/161 xiv, 32, 148, 303
  Directive 2001/83/EC xiv, 32, 148, 303
  distributors 55, 315
  documentation 39
  equivalent identifiers 40, 313
  expiry dates 32, 308
  falsified medicinal products 32, 148, 310
  fraud 41, 46
  free samples 55
  hub repositories system 47, 49, 313
  importation 55, 315
  manufacturers 39, 310
  Marketing Authorisation 55, 305, 315
  national repositories 47, 313
  parallel distributors 55, 315
  parallel importers 55, 315
  partial package supplies 45
  personnel xv, 43, 311
  printing quality 36, 307
  product codes 32
  recalls 55, 315
  record keeping 39
  removing/replacing 40, 55
  repositories system 46, 312
  safety features xiv, 32, 148, 303
  serial numbers 32, 308, 313
  status reversals 39
  stolen products 55
  supply chains 42
  supranational repositories 47, 313
  tampering xv, 32, 41, 43, 303
  technical specifications 34, 306
  verification 38, 309
  wholesale distribution 310
  withdrawn products 55, 315
United Kingdom (UK)
  see also Human Medicines Regulations 2012
  guidance
    active substances/APIs 240, 241
    brokering/brokers 206
    distribution 240
    exports 122, 124, 145
    falsified medicinal products 123, 145, 171, 180
    GDP 115
    imports 145, 153, 173, 240
    manufacture 240, 241
    wholesale dealer's licences 114
    wholesale distribution 113, 123, 145, 171, 180
  imports 145, 153, 173, 240, 289
  legislation
    active substances 219
    brokering/brokers xx, 196
    distribution/manufacture/importation 219
    GDP 220, 222
    glossary of xii
    GMP 220, 222

preface overviews xx
TSE 23
wholesale distribution xx, 64
licensing imports/exports 289
scientific advice 7
wholesale distribution xx, 64, 113, 289
United Nations (UN) 241
unlicenced medicinal products
see also special medicinal products
centrally authorised products 178
DMRC 179
EMA 178
equivalent licenced product availability 179
exports to non-EEA countries 148
Human Medicines Regulations 2012 178, 179
importation 12, 178
Marketing Authorisation 178
MHRA 178
TSE 23
wholesale distribution 148, 177
unlicenced sites, returns 165
unusual sale patterns 110
uploading information 47
usage of language 203

V

vaccines 144
validation considerations 91, 107, 239, 233, 297
vehicles 102
verification considerations 38, 309, 310
veterinary care professionals 139
Veterinary Medicines Regulations 2013 144
veterinary products 11, 115, 144, 261

W

waivers 245, 249, 252, 300
walk-in cold rooms 160
washrooms 90
weather conditions 89, 109, 232
websites 19, 143, 261
white lists 249
WHO *see* World Health Organization
wholesale dealer's licences
see also wholesale distribution
application procedures 70, 114
brokering/brokers 123
certificate of registration 121, 125
competent authorities 173
conditions for 72, 121
Data Processing Group/MHRA 11, 12
definitions 68
Delegated Regulation (EU) 2016/161 xv
Directive 2001/83/EC xv, 25, 128, 145
distribution 121
documentation 73, 76, 122, 124
duration of licences 72
EEA States 74, 122, 124
exempt ATMPs 80, 127
exempts from requirements 69
exports 25, 75, 122, 124
fees 74, 115
financial introductions from non-EEA countries 147
GDP 76, 115, 172
granting licences 68, 115, 172
handling procedures 121
holding conditions 72, 121
Human Medicines Regulations 2012 130, 178
human use medicinal products 115
importation 75, 122, 126, 146
licence holding conditions 123
manufacturer's licences 123, 148, 153
non-EEA countries 25, 124, 146
obligations of 72
personnel 73, 121
physical introductions from non-EEA countries 147
refusals 71, 115
regulatory action 173
Responsible Person(s) 77, 121, 125, 130
revoking licences 74, 123, 127
safety features xv, 75
Schedule 6 Licences 80
specified person requirements 75
standard provisions of 72, 78, 125, 127
suspensions 74, 123, 127
Title VII of Directive 2001/83/EC 25, 128, 145
UK guidance 114
unlicenced medicinal products 12
variations to 115, 123
veterinary products 115
withdrawing licences 127
wholesale distribution
see also brokering; wholesale dealer's licences
ACMD 143
adverse reactions 79, 127, 180
advertisements 80, 126, 167
ambient products 158, 161, 165
Article 126a authorisation 65, 121, 125
authorisations 65, 154
BGMA 177
blood-derived products 30, 138, 144
CAPA 136
centrally authorised products 178
change control 135
cold chain products 156, 158
communication 66
Community Code changes xiv
complaints 97, 111, 236
compliance 28, 120, 169
computerised systems 91

contacting activities 85, 100
continued supply 73, 175
controlled drugs 138, 147, 148, 155
controls 90, 137
customer qualification(s) 168, 170
customs procedures 147, 149
data integrity 185
decommissioning 310
defective products 179, 180
deficiencies 117
Delegated Regulation (EU) 2016/161 xv, 41
destruction procedures 96, 137
deviation management 136
Directive 2001/83/EC 25, 65, 128, 145, 177, 193
Directive 2011/62/EU 25, 145
discontinued products 177
diverted medicines 173
documentation 28, 29, 73, 76, 92, 122, 124, 149
Drug Alerts 181
drug misuse 139, 143, 155
due diligence 127, 169, 171
EEA States 66, 74
electronic communication 66
EMA 74, 125
emergency plans 28, 73, 122
environmental controls 90
equipment 73, 88, 90, 108, 121, 232
EU legislation xx, 23
exports 66, 75, 122, 124, 145, 289
falsified medicinal products 66, 75, 123, 145, 148, 171, 180
fees 66, 74, 115
flammable products 137
fraud 173
freezers 160
free zones/free warehouses 84, 107, 145, 149
GDP 76, 83, 115, 137, 172
General Sale List 268
GMP 153
herbal products/substances 66, 121, 125
holding authorisations 154
Home Office Drug Licensing and Compliance Unit 138
homeopathic medicinal products 67
Human Medicines Regulations 2012 64, 130, 178, 267
immunological medicinal products 30
importation 67, 75, 122, 126, 145, 146, 153, 173, 178, 289
IMPs 68
Incoterms® 2010 rules 151
information provision 125
inspections/inspectors 67, 115
insulated boxes 104
introduced medicinal products 146
ketamine reclassification 142
labelling 103
large refrigerators 160
legislation 23, 64
lethal injection drug controls 141
licences/licensing 67, 68, 114, 289
lisdexamfetamine controls 142
management 85, 136
mandatory requisition forms 139
Manufacturer's/Importer's Authorisation 153
manufacturer's specials licences 148, 153
Manufacturing Authorisation 26, 153
Marketing Authorisation 66, 73, 75, 121, 124, 178
MHRA 114, 125
misuse of drugs 139, 143, 155
name consideration 80, 125, 226
narcotics 30, 103
non-EEA countries 124, 145, 145, 146, 150
outsourced activities 85, 100
parallel distribution 175
parallel importation 173
personnel 26, 73, 86, 121, 267
pharmaceutical refrigerators 158
pharmacy medicine guidelines 268
planning inspections 115
precursor chemical controls 138
preface overviews xix
premises 73, 121, 125
procurement/procuring 144, 145, 154
product recalls 174, 180
products requiring special conditions 103
product withdrawals 180
psychotropic substances 30, 103
qualification(s) 168, 170
Qualified Person(s) 148
quality defects 180
Quality Management 135
Quality Risk Management 135, 163
quality systems 28, 75, 84, 123, 135
radioactive materials 89, 103
radiopharmaceuticals 30, 68, 137, 144
raw materials 127
recalls 174, 180
receipts 95, 232
records 28, 74, 122, 124
refrigerators 158
registrations 69
Regulation (EC) No 726/2004 25, 67
Regulation (EC) No 1234/2008 67
Regulation (EC) No 1394/2007 67
Responsible Person(s) 136, 137
retail sale or supply 124
returns 164, 165
risk-based inspection 118
risk management 28, 135, 163
risk rating 120
safety xv, 41, 75
sales representative samples 167

Y

Z